Real-Time Software Design

Philip Heller

Real-Time Software Design

A Guide for Microprocessor Systems

With 85 Illustrations

Birkhäuser
Boston · Basel · Stuttgart

Library of Congress Cataloging in Publication Data
Heller, Philip
Real-time software design.
1. Real-time data processing. 2.Microprocessors—
Programming. I. Title.
QA76.54.H44 1986 001.64'404 84-9331

CIP-Kurztitelaufnahme der Deutschen Bibliothek
Heller, Philip:
Real-time software design / Philip Heller.—Boston ;
Basel ; Stuttgart : Birkhäuser, 1986.
ISBN 3-7643-3201-8 (Basel . . .)
ISBN 0-8176-3201-8 (Boston)

ISBN 0-8176-3201-8
ISBN 3-7643-3201-8

Typeset by David E. Seham Associates, Inc., Metuchen, New Jersey.
Printed and bound by R.R. Donnelley & Sons, Harrisonburg, Virginia.
Printed in the U.S.A.

9 8 7 6 5 4 3 2 1

This book is dedicated to three of today's outstanding creative people:
to my mother, to my brother, and especially . . .

To my wife, LISA

I am deeply indebted to the following friends, who have played a part in the writing of this book:

Bob Blum
Alan Finkel
Clarissa McDaniel
Jeff McDaniel
Constance Reid
Dave Wigginton

Preface

Computers these days spend a fairly low fraction of their time computing. In fact, the very word "computer" has become something of a misnomer.

In the American History museum of the Smithsonian Institute in Washington, D.C., there is an exhibit of early computers. Three features of these machines are striking. First, they are enormous, especially in comparison to their capabilities. The museum visitor who has just come from the Natural History building next door may be reminded of fossilized dinosaur bones. Second, they don't look at all like modern computing machines. The cases are made of crude metal or beautifully worked wood, recalling an approach to the design of scientific apparatus which belongs to a previous generation. Lastly, the function of these machines is mainly to compute— to perform rapid arithmetic.

The computer of today bears little resemblance in size, form, or function to its ancestors. It is, most obviously, smaller by several orders of magnitude. Its form has changed from the carefully crafted one-of-a-kind instrument to the mass-produced microchip. But the change in its function is perhaps the most dramatic of all. Instead of being a computing engine, it is a machine for the processing of information. The word "processor" has come into common usage. A processor used to be a central processing unit—a set of wires and vacuum tubes, or later a set of printed circuit boards—which was nestled deep within the computer. Today a processor is an off-the-shelf component.

These changes in size, form, and function are the reflections of the evolution of the tasks to which we apply computers or processors. In this book we examine one family in the new population of applications, namely real-time devices.

The purpose of this book is to teach some valuable techniques to the engineer or engineering student who is faced with, or anticipates being faced with, a real-time programming assignment. The goal is to give the reader some tools to make the project more enjoyable and more successful. We will therefore confine our discussion to the practical aspects of our topic. For example, we begin with a description of real-time software, rather than a formal definition. From there we develop a motivation for studying this topic. This motivation can be briefly summarized as follows: if you aren't familiar with the behavior of real-time systems, then your program may well end up designing itself, rather than surrendering that control to you. This situation is characterized by high stress, low enjoyment, and very little success.

Having justified the importance of our topic, we examine what we call the "hardware substrate" of all real-time systems—the devices with which the software interacts. The most important of these is the processor. We pick as an example the Intel 8Ø86, which is the heart of the IBM Personal Computer. The 8Ø86 family of microprocessors has been steadily expanding, and now includes the 8Ø88 (a slightly simplified 8Ø86), and some enhanced versions such as the 8Ø286 and the 8Ø386. For our investigation we have restricted ourselves to the more straightforward 8Ø86, even though its more powerful relatives are almost certainly preferable hardware design options. The reason for this is that we want to write about software design skills, not about how to write code for any single processor, and by selecting the 8Ø86 we minimize the effort the reader has to spend in learning the instruction set.

Once the hardware basis has been established, we are ready to discuss design techniques. As in any other aspect of life, you can only make use of a concept if you can communicate it in some sort of language. We therefore provide the reader with four languages for expressing software design ideas. The vocabulary of these languages is pictures, not words. An idea expressed in such a language is a diagram, not a paragraph. The four languages we examine are flowcharts, structure charts, data flow diagrams, and state diagrams. Each is appropriate for expressing a different sort of idea. We will examine each in turn, discussing its advantages and shortcomings. We will conclude that the data flow diagram is the most appropriate language for real-time software design, although the other three are certainly of value.

In Chapter 4 we look at some techniques for turning a data flow diagram—a design—into reality—a program. Chapter 5 applies these techniques to a practical example.

In Chapter 6 we briefly look at some techniques for optimizing system performance.

It is our hope that the reader will learn some skills which will be of practical value. Whenever possible we avoid purely theoretical discussions or formal definitions. Instead, we try to teach some realistic methodologies which will make the reader's design tasks more fulfilling.

Mountain View, California
October, 1986

Contents

CHAPTER 1
What Is Real-Time Program Design?

1.1 What Is Real Time?

Lay people sometimes find the term "real time" disturbing. It implies an equal and opposite "unreal time."

In the context of computers, a real-time program is one which runs continuously, reacting to spontaneously changing inputs. For programs, the opposite of "real time" is "batch": a batch program's inputs are predetermined and prepackaged before the program is run.

Let's clarify the distinction with some examples. An example of a batch program would be the kind of payroll program used by most large companies. Such a program is typically run on a large mainframe computer system. Every pay period, attendance information from the employee's time cards is entered into a file—in practice, into a disk file. Then a program is run which collects salary information from another disk file, performs the appropriate computations for gross pay, withholding, and accrued benefits, and prints a check and a check stub. What is noticeable about this system is its high degree of isolation from its environment. The input data is neatly prepackaged into records and files. The output character strings for the checks and stubs need only be delivered as parameters to a system utility subroutine; the moment-to-moment fine control of the printer is all taken care of. And once the output data is delivered to the printer, the social system takes responsibility for the rest of the dirty work, such as delivering the checks to the employees.

As an example of a real-time program, consider a program to control a portable microprocessor-based electrocardiogram machine. Such an instrument might have sensors attaching to the patient's chest, a front panel for issuing commands to the system, and a small thermal printer for drawing a hard-copy graph of the patient's heartbeat. The program for this machine must do much more than process heartbeat data. It must gather its own inputs by monitoring the front panel and the sensors at sufficiently brief intervals. It must interpret keystrokes at the front panel into high-level commands. Lastly, it must synchronize its outputs with the printer, converting the desired graph into low-level printer command signals. This program seems to be much more involved in its environment than the program in the previous example. It is this aspect of involvement and interaction with the environment which makes the design and implementation of real-time programs exciting. They run not only in "real time" but also in "real space," right here in the "real world."

What are some of the characteristics of real-time programs and the systems they run in?

First, because a typical real-time instrument is so connected to its environment, real-time systems tend to be designed for specific uses. They possess special purpose peripheral devices such as sensors and drivers, along with the digital interface hardware which lets them communicate with these devices. This specificity makes them well adapted to a particular task, but (therefore) not very adaptable to tasks in general.

Contrast this to the mainframe computer of our batch program example. Such a system usually has very general purpose peripherals: disks, tapes, printers, and terminals. The payroll program would reside on disk and be copied to RAM by the operating system just before execution. This computer could as easily load and execute an inventory program, a compiler, a statistical analysis social-sciences package, or a game. On the other hand, real-time systems, and especially microprocessor-based real-time systems, execute their programs from ROM. There is only one program, so there is no need to keep it on a mass-storage medium and retrieve it into read/write memory.

This special-purpose nature is the reason for the great diversity of real-time products. Microprocessors executing real-time code can be found in terminals, robots, medical instruments, microwave ovens, communication devices for the handicapped, American cruise missiles, Soviet SS-20 missiles, French Exocet missiles, and video games.

A second characteristic of real-time programs is that data is very thoroughly transformed. Batch programs read nicely formatted data and produce nicely formatted data. Real-time systems have no such luxury: they have to gather and format their own inputs, then do data processing, then break down the output formats into low-level output control signals. The number of times a unit of data is passed as a parameter to some subroutine for processing or refinement tends to be higher in real-time systems.

A third characteristic is that real-time systems appear to do many things at the same time. When writing software to run on a general-purpose computer system, we conceive that only one routine of our program will execute at a time, and within that routine only one instruction at a time, whether the program is written in FORTRAN, C, LISP, Pascal, or BASIC. Real-time systems, on the other hand, appear to be executing several or many routines at once. The electrocardiogram machine described above, for example, gives the impression of simultaneously executing four tasks: the sensor scanner, the front panel scanner, the data processing, and the printer driver. This is, of course, an illusion: the illusion of multiple concurrent tasks. It is one of the most powerful illusions of computer science, so for the benefit of all aspiring computer illusionists we will study it in detail in Chapter 4.

A fourth characteristic is feedback. A real-time system can be designed so that there are inputs to monitor the effects of the outputs. This makes it possible, for instance, for a system to assert an output signal for precisely the correct length of time, even though that duration is not known to the programmer at the time of writing the program. The correct length of time may even vary from situation to situation, from execution to execution of the program.

Let's look at an example of feedback. Consider a microprocessor-based controller for an automatic dishwashing machine. One task of the processor's program is to assert an output signal which opens a valve to allow hot water to fill the dishwasher. When the tub is full, the program has to assert a second output signal which closes the valve again. This can be implemented with or without feedback, as the two flowcharts in Fig. 1.1 represent.

The flowchart on the left solves the problem without feedback. The

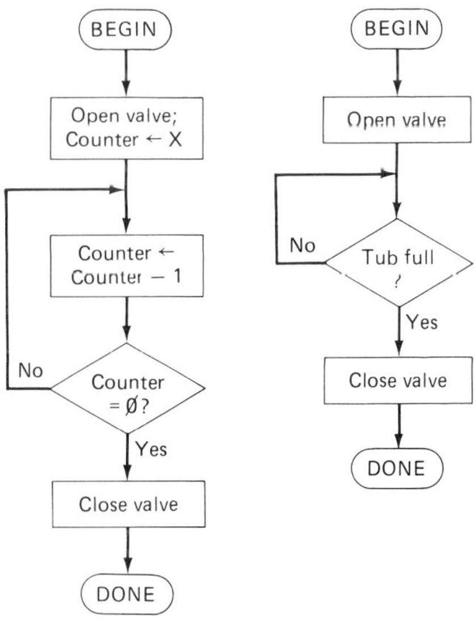

Figure 1.1

value X which is intially loaded into the counter is carefully calculated from the amount of time it takes to fill the tub and the amount of time it takes the processor to execute a single iteration of the loop. The flowchart on the right uses feedback. The "TUB FULL?" query in the decision diamond implies that there is a sensing device which tells the processor when the tub is full. This device makes the product more complex and more expensive, but it has distinct advantages. The feedback implementation is considerably more reliable because it does not depend on the accuracy of the calculation of X. In fact, this calculation is not possible, because the ideal value for X will fluctuate from day to day. Changes in water pressure will affect the amount of time it takes to fill the tub. More insidiously, the clock circuit which provides timing to the processor, thereby determining how long it takes to execute the loop, may be affected by heat or other environmental forces. If the washer is being mass produced, there are additional practical considerations. A cheaper clock circuit with a somewhat different period may become available. The company could stand to save a lot of money if the cheaper circuit could be used in the product without requiring a software change. In a commercial environment, great savings can be realized from a small reduction in the manufacturing price of a product, but software changes can be very costly (particularly if large quantities of obsolete ROM's containing obsolete code must be scrapped).

Even systems with no explicit feedback can have an implicit sort of feedback: the operator can personally "close the feedback loop" by stimulating an input in reaction to an output. This could happen, for example, if the operator of our electrocardiogram noticed a very strange waveform coming off the printer and issued a command to the instrument to abort the current print operation and remeasure the patient's heartbeat.

The essential requirement of any real-time system is that it has to respond quickly and correctly to its environment. What a requirement! It's difficult enough for *us* to respond quickly and correctly to our day-to-day lives; how can we be expected to program machines to do the same?

This is a vague question, and we will not offer a quick answer here. This entire book is a partial answer. At this point let us just observe that any solution to this general question must be greatly aided by FAST PROCESSING and ABUNDANT MEMORY. Integrated circuit manufacturers are obligingly addressing this need by constantly upgrading the performance of processors, memory chips, and peripherals. They do this because they realize that no matter what improvements they make, companies, universities, and institutes will always invent new applications to take advantage of the improvements.

From these observations we can draw two conclusions which are relevant to our careers (assuming that the reader is either an engineer or a future engineer).

First, the evolution of hardware components for real-time systems and of software requirements for those systems is not going to slow down in the foreseeable future. In 1980, the majority of real-time products were controlled by eight-bit microprocessors, of which the Z-8Ø was the commonest and most sophisticated. By 1984, sixteen-bit processors were on the rise. Of the common ones, the simplest was the 8Ø88 and the most complicated was the 68ØØØ. Digital peripheral devices have similarly evolved; video controllers for CRT display terminals, for example, became available on chips from a variety of manufacturers. The lesson in this for all of us is that it is not enough to master any particular processor or technology. In order to be of ongoing value to the engineering community, engineers have to learn how to learn processors and technologies.

The second conclusion we can draw from industry's commitment to the evolution of hardware is this: real-time programming will be with us for a long, long time.

1.2 What Is Programming?

We can define the programming of processor-based real-time systems to be the providing and implementing of an algorithm for generating the correct outputs at the correct moments, given any combination of inputs and their timing relationships.

Like any broad definition, this one is of little value (everyone knows what programming is!) until we discover its practical implications.

The first part of our definition concerns providing an algorithm. To whom or to what do we provide it? First we provide it to ourselves, in the sense of presenting it to our consciousness—thinking about it—so that we can translate it into a computer language. Once the computer-language code is written, and entered into a computer system as a file, it is provided to an assembler or to a compiler for translation into the appropriate processor's machine language. The machine language algorithm is (somehow) loaded into a prototype of the system it is to control, thus being provided to the product's processor to serve as its program. Lastly, when the program is completely developed and debugged it is provided to an archive system and, in a commercial environment, to the manufacturing department for mass production. The engineer is the one who does all of this providing, sometimes in an electronic context, sometimes in a social context, sometimes in a mental context. There's more to the job than just writing code.

The next part of our definition of programming concerns generating the correct outputs. This means providing just those output signals which will make the system perform in the desired way. The desired performance may be specified precisely, but the actual product will almost certainly not perform exactly according to specification, unless the specification is very simple or very vague. The actual performance can be quite close, hopefully will be within an acceptable tolerance, but it usually will not hit the exact bull's-eye because such exact control over the fine behavior of a sophisticated program is nearly impossible. For example, if the specification for a video display terminal calls for it to process 19,200 characters per second, then it should be considered acceptable if it processes 18,950 characters per second, plus or minus 500 characters.

This nondeterministic or "fuzzy" aspect of the definition of correct outputs shows up especially when we consider the requirement that the outputs be correct at the correct moments. Transitions of output signals cannot be planned to within a time tolerance which is smaller than approximately the order of magnitude of the period of the processor's clock. Within the processor an output signal can only change on a particular clock transition during the execution of an OUTPUT instruction. Moreover, there is a certain uncertainty in the amount of processing time between the stimulus which causes the output transition and the transition itself. Tolerances cannot be met if they are smaller than approximately this processing time.

The last part of our definition requires that the outputs be provided given any combination of inputs and their timing relationships. Inputs can behavein three ways: normally, abnormally, and catastrophically. Behavior is normal when all the inputs are acting within any limits beyond which they are not reasonably expected to go, and beyond which the program or the system is not guaranteed to function. Due to unforseen circumstances the inputs may not behave as expected. For instance, an input signal may make more transitions per second than expected. This is abnormal behavior; it is a design bonus if the system can stay calm during the abnormality and be able to resume ordinary operations once normalcy is restored. It is particularly good if during the abnormality the system takes no action which might be destructive to data, code, hardware, or life. The last kind of behavior is catastrophic, for instance when power fails (from an analog designer's point of view power is after all an input) or when a three-thousand volt signal appears at the input to a TTL gate. The best you can do here is to take precautions against those catastrophes which are likely to happen or have been observed to happen in similar circumstances.

1.3 What Is Design?

There is a tendency in computer literature—particularly in textbooks and technical manuals—to give definitions which are correct and elegant but are difficult to translate into day-to-day experience. Since the goal of this book is to communicate practical design skills, we will avoid the temptation to provide a really brilliant thumbnail definition of design. We will instead be content to give a casual definition within the context of real-time software design.

The word "design" is both a noun and a verb. As a noun, we will take it to mean a description in some high-level language of the program to be written. As a verb it means the process of creating the description. (We will also use "design" as an adjective, as in "design skills" above, or "design technique," "design tools," or "design flaw." Computers have done more for the propagation of the adjective than any other industry in history. If you're writing about computers, you can make an adjective out of ANYTHING, especially nouns. To find an example of this phenomenon you need look no further than the cover of this book.)

The design is generated in a high-level language because its purpose is to convey a feeling for the intent and the nature of the program in a concise way. The term "high-level" when applied to formal language has two connotations. First, there is the idea of being close to the designer's native language relative to the programming language or to other formal languages. A design for a program should of course be executed in a more nearly natural language than the program it represents; otherwise the program would be more useful than the design. (This sometimes happens. If the design is generated as an afterthought when the program has been completed, then the programmer probably found the design language more cumbersome as a design tool than the programming language.) The other sense of "high-level" is "powerful." In this sense FORTRAN is on a higher level than Assembly Language because a single FORTRAN statement may translate into many Assembly Language statements. These two connotations, naturalness and power, are not equivalent. LISP is definitely more powerful than BASIC: it could take all day to write a BASIC program to emulate the function of a few intricate lines of LISP code. But BASIC more closely resembles English than LISP does.

What languages are typically used for real-time software design? Besides natural human language, the four common formal languages are FLOWCHARTS, STRUCTURE CHARTS, STATE DIAGRAMS, and DATA FLOW DIAGRAMS. They all happen to be diagrammatic—that is, they are based on diagrams rather than on prose alone. All have inherent advantages and all four have inherent disadvantages. In this book we reveal a strong prejudice for data flow diagrams, although we will also pay attention to the others.

As a medium for design, natural language has the distinct advantage of being immediately understandable by practically anybody who wants to learn about the design. The only instance in which this would not be the case would be an international organization where the design is transferred across a language border. The problem is that natural language is not nearly as "dense" as the synthetic diagrammatic languages, so a natural language design is much more lengthy than a diagram. One purpose of a design is to explain the program to some one who hasn't seen it before. If the design is so lengthy that the newcomer is discouraged from learning about the program, then the design has not fulfilled its purpose.

There is another disadvantage to designing in natural language. It stems from the nature of the spoken and written word. People read designs with the eye and the brain. When looking at a diagram, the eye can jump from block to block, scanning in two dimensions, picking up meaningful information wherever it looks. This is the natural function toward which our visual apparatus has evolved: we are used to scanning planes. This is not what happens when we read a prose composition, which is essentially linear: in order to understand an arbitrary sentence, it is generally necessary to have understood all of the previous sentences. Thus a natural language description requires a disciplined approach, while a diagrammatic design can be assimilated much more casually.

A last disadvantage stems from the fact that in any design some information is more important than other information, and some information is of a different nature from other information. In a written paragraph, all words are at first glance just like all the other words; it is only when you read the words in context that meaning emerges. Diagrams are typically composed of boxes, arrows, and labels. It is immediately obvious even to some one who is not familiar with the language of the diagram that an arrow conveys a different sort of message from a label, and that all boxes convey the same sort of message. Thus a diagram can be easier to learn than a document because it inherently structures information.

The formal languages which we will discuss in Chapter 3 are all diagrammatic. They can all be drawn with boxes, bubbles, and arrows, and templates for all the symbols are available at most stationery stores or college bookstores. The main thing to remember with all of them is that it is more important to convey clear information than it is to stick to the rigorous definition of the language. Each of the four formal diagrammatic languages of Chapter 3 conveys a unique distillation of the nature of the system being described. Each has its own point of view. Flowcharts communicate the order of execution of the different parts of the program; structure charts depict the hierarchy of subroutines; state diagrams show how the system changes as its inputs change; and data flow diagrams show the paths information takes and the transformations it experiences. None of these points of view alone may be completely appropriate to a particular design problem. Designing is a creative process, and you might have to get creative with the formal language in order to convey an idea in just the right way. After all, the people who write technical documents make up new words and nuances within the natural language every day. Why shouldn't you have the same prerogative with a formal design language? If your company or department has the attitude that you must rigorously conform to a given definition of a given formal language without ever getting to take liberties with it, then you are probably too creative to be working there.

Why is it necessary to design software? For many problems there is one obviously best way to start the program, one obviously best way to proceed, and one obviously best way to finish. You sit down and write the program, and that's that.

In fact, this approach works as long as the program does not have to be too complex.

At this point we should carefully define complexity, because there is some room here for confusion. In the study of algorithms, the term *computational complexity* has come to mean the amount of time and memory required by an algorithm, as a function of the number of input parameters. This is *not* the sense in which we use the word, which we take to mean *an informal measure of the intricacy, sophistication, and subtlety* of a program.

With this definition in mind, we can identify five major factors which contribute to program complexity. They are depth of hierarchy, data path lengths, program length, reaction time, and self-reference.

Depth of hierarchy refers to the number of levels of subroutine calls in the program. It contributes to complexity by making the highest-level routines extremely powerful, having so many subordinate routines under their control. Not just software systems but systems in general which are composed of many hierarchical levels tend to manifest surprising traits.

Data path lengths also contribute to complexity. A data path is the route which data follows while being passed as a parameter from routine to routine, being processed and refined along the way. The path length is not a physical measure; it is the number of times the data is passed from one routine or process to another. Data is hardly ever passed for no reason; it is passed so that it can be processed, so the data path length is a measure of the amount of refinement which data experiences.

The third factor is reaction time. This is a measure of how speedily a program reacts to changes in its environment, relative to the processor's speed. This would seem to be a quantitative factor, unrelated to the qualitative nature of complexity, and indeed this is true for long reaction times. But when the reaction time approaches the order of magnitude of the processing speed of the processor, the program must become more and more intricate in order to be able to react in time. For example the program to control a data terminal which has to receive and process 11 characters per second (standard teletype speed) can afford to be much more elegant than one which has to receive and process 19,200 characters per second (the high end of modern asynchronous data transmission speeds). The faster terminal has to cut corners in order to keep up with its input stream, so its routines and data structures will tend to be interrelated in more twisted, tortuous ways.

The fourth factor is simply program length. At first glance program length, like reaction time, appears to be a quantitative factor with little relevance to the quality of complexity. It may seem naive to think that length on its own could enhance complexity, but to my experience there is no such thing as length alone. Programs which are long are never just long; their very length requires a supporting structure, in the same way that a long bridge requires supporting arches or towers, and such structure is part of complexity.

The last factor is self-reference. This is the degree to which the program processes data which is not so much some refinement of its physical inputs as much as a representation of the state or performance of the program itself. In other words, self-reference is the degree to which the program controls and monitors itself. The simplest and least complicating kind of software self-reference is the recursive subroutine, which repeatedly calls on itself to operate on data which (hopefully) will eventually be transformed into a value which will terminate the self-call. Recursion makes a program mode complex by blurring the otherwise clear distinction between levels of hierarchy, and by creating variable (hopefully bounded) data path lengths. As a familiar example of recursion, consider a subroutine to calculate the factorial of its input parameter. Call the input parameter "X". A recursive implementation of this function would return a value of 1 if X was Ø; otherwise it would call itself to calculate the factorial of X – 1, and multiply the result by X. The depth of hierarchy of this subroutine, in other words the number of subroutine calls it makes, is not fixed; rather it varies from execution to execution of the subroutine. A more complicating type of self-reference could be found in a portion of an operating system which dynamically allocates blocks of RAM to requesting programs, based on the availability of functioning unused memory. With such a relocatable system, a program in execution could reside anywhere in RAM space. If the allocator itself executes out of dynamically allocated memory, we have a self-reference by virtue of the allocator's allocating itself. The result is a more complicated system, since there is no way of predicting where a program will end up.

(An insidious kind of self-referential allocation is used by a family of prank programs called "tapeworms." A tapeworm makes a copy of itself somewhere in memory besides where it is executing, erases the original, and jumps to the copy. By the time you've noticed it, it has gone somewhere else. In a "soft network" of computers loosely connected by modems, a tapeworm can call another system on the network and jump through the telephone to a different system. The worst kind of network tapeworm makes two copies of itself, presenting the network with a population explosion. At any rate, tapeworms have a profound ability to complicate the systems they reside in, and this complexity comes from their self-referential nature.)

These five factors are not the only elements which contribute to complexity, but they should serve to give you a feeling for the kinds of things that complicate programs.

Real-time programs are required to be ever more complex. This tendency is maintained by two inexorable forces: hardware advances and market competition.

Hardware advances, as we will see in the next chapter, have provided the physical substrate to support all of our five components of complexity, and market competition has provided the incentive to exploit the substrate to its limits. For example, faster processors permit faster program reaction times, which we have identified as one factor of complexity.

This exploitation in turn creates an incentive for hardware manufacturers to enhance the substrate by producing denser, faster, cooler, cheaper, and occasionally easier to use hardware devices. Enhancements to the substrate encourage companies which manufacture systems to develop more complex products, and so on in what is more of a tornado than a spiral. The effect this has on the engineer is predictable and ubiquitous: every major project is more complex than the last.

Design is the one thing which makes very complex real-time software possible, by providing an organized structure to hang the complexity on. Why does there have to be any structure at all to support the complexity? Why can't the complexity just stand on its own?

These questions are both technical and philosophical in nature, and so there are no purely technical answers. General Systems Theory is a discipline which investigates systems of all kinds, combining rigorous and nonrigorous thought processes. Some recent advances by general systems theorists are both surprising and relevant to the subject of software system complexity. (See JANTSCH.)

Imagine a system—any kind of system—which processes and transforms a commodity. The system could be a real-time program and the commodity could be the information flowing through the data paths. Or the system could be a population of plants in a field and the commodity could be energy, which enters the system as sunlight falling on the leaves and is transformed by photosynthesis to chemical energy in the form of sugar. The system could even be a high-pressure weather system, with the commodity being a mass of water molecules evaporating off of a lake below.

Whatever the system and whatever the commodity may be, the system's *throughput* is the rate at which the commodity is processed by the system.

As the amount of the commodity to be processed increases, the system may eventually reach a point of crisis, where it can no longer tolerate an increase in throughput. Now comes the startling observation of General Systems Theory: at this crisis point, the system will evolve dramatically, spontaneously reorganizing itself into a more complex configuration which is able to handle the new throughput requirements. If this self-organization is not catastrophic—that is if it has not broken—the system can experience a series of reorganizations in response to further throughput increases. Let's look at some examples. (The first is from JANTSCH.)

When a water tap is turned on low, the stream of water coming out of the faucet is cylindrical and smooth, exhibiting laminar flow. The water molecules all travel downward under the influence of gravity and water pressure, each molecule behaving much like the others. The system is the faucet and the jet of water, the commodity is the water itself, and the throughput is the rate of flow. This system is not very complex or very highly organized. As the tap is opened wider, thereby increasing the throughput demands on the system, a beautiful thing happens. The cylindrical symmetry of the stream is suddenly broken and turbulent flow is manifested. The stream develops graceful twists and ridges. The turbulent system is more highly organized than its laminar ancestor. In the laminar system there is no intermediate level of structure between the molecular level and the system level; in the turbulent system the molecules are organized into streams which in turn compose the jet of water. It is precisely these complex strandlike streams which enable the turbulent system to accommodate the increased water flow which we imposed on it when we opened the tap wider.

As a second example, consider a large oceanic body of water which gradually heats up during the spring and summer and cools off during late summer and fall. In the tropics, the water can absorb quite a lot of heat with no opportunity to dissipate it for many months. Later in the season the atmosphere above the ocean cools enough to be able to accept heat from the water. At first this heat is spread through the air by conduction, with individual air molecules transferring thermal energy to their neighbors by collision. As the amount of heat given to the atmosphere increases, conduction alone is no longer sufficient to dissipate heat from the region of air immediately above the warm water. At some point the method of dissipation switches to convection, in which the air molecules receive heat from the water and then, instead of passing the thermal energy along, migrate with it away from the region. Thus an intermediate level of structure is formed, consisting of the pathways or channels which the molecules follow: winds. If even this method is insufficient the winds themselves may become more highly organized, into hurricanes.

We see that when physical systems are subjected to increasing throughput requirements, they can spontaneously manifest new modes of organization and new properties. Let us apply the term "emergent behavior" to the system's new way of existence. Can we apply the principle of self-organization to computer programs? Do programs ever manifest emergent behavior?

I believe that we can, and that they do. We can take throughput to be the amount of data to be processed, the speed at which it must be processed, and the sophistication of the processing. What then is the emergent behavior? It is simply observable modes of behavior which were not deliberately designed into the program and which do not violate the program's requirements (otherwise they would be bugs, not emergent behavior). Although unexpected behavior is readily seen in most developing complex programs, I can think of no more clear-cut example than the emergent behavior manifested by the ever-popular PAC-MAN video game.

I doubt if any reader is unfamiliar with PAC-MAN, but I will briefly summarize it just in case. The player controls the movement of a voracious yellow disk named PAC-MAN as it travels through a maze displayed on a color graphics screen. The maze is floored with small dots which appear to be food for the disk, since munching noises are heard and the dots disappear when the disk passes over them. Four brightly colored ghostly villains lurk in the maze, chasing after PAC-MAN with violent intent. If they touch PAC-MAN he expires; he can turn the tables on them, however, by eating one of four larger dots which temporarily give *him* the power to overcome *them*. From time to time valuable prizes appear in the maze (valuable to a PAC-MAN anyway, food mostly); if PAC-MAN eats them the player receives many points. When all of the dots are eaten up the maze is refilled and PAC-MAN and the player get to start again. With each new maze the ghostly villains move faster and are harder to vanquish, and the valuable prizes are worth more (very many more) points.

So the strategy for each new maze is to avoid ghostly villains long enough to eat all the dots and, hopefully, the prize. It turns out that each successive maze has a "pattern"—that is, a fixed and constant path which if followed is practically guaranteed to put PAC-MAN in the right place at the right time to acquire as many points as possible without ever touching the ghosts. Each maze has a solution. This is surprising. In fact, it is emergent behavior. The patterns were never programmed into the game. (Would you buy a video game for your bar which deliberately let novice players play for hours on end for twenty-five cents? Neither would I.) The only behavior which was programmed in was the way the ghosts and the prizes behaved. These algorithms are a level below the level of the patterns. Shortly after the patterns were discovered, two books about how to become invincible at PAC-MAN appeared on the mass market. Games which were variations on the original theme appeared, partly in an effort to stay one step ahead of the pattern finders.

Not all complex programs demonstrate such dramatic emergent behavior, but any unplanned emergent behavior in a program in any degree is likely to be undesirable. If a program is destined to have behavior and organization, by all means let it be the behavior and the organization that you want it to have, not what it comes up with on its own.

We can see how truly beneficial a well-organized supporting structure can be by looking at some examples from outside the realm of computer science. It would be very difficult to erect a useful building taller than about ten stories without a hard rigid frame, but by first setting up a skeleton of structural steel it is possible to build a skyscraper more than one hundred stories tall. In the realm of prose, it is almost impossible to imagine writing a book (particularly this book!) without following a detailed outline. The most dramatic example comes from biology: we cannot conceive of a group of atoms of carbon, hydrogen, oxygen, sulfur, and various other elements spontaneously combining to form a human body. As a random event this could not happen once in a billion years, but thanks to the genetic organizational structure provided by DNA, a billion new human beings come into existence every generation.

The above examples show how structure permits complexity. In engineering, it is deliberate design which provides that structure.

In developing a large program, when should one design and when should one implement?

It is tempting to try to complete the design before beginning the implementation. From a managerial standpoint this is definitely the securest way to proceed, and it is cause for celebration when a project is successfully developed this way. In practice, however, there are two strong forces which compel the design to change after implementation has begun. The first force is imposed when management decides to change the specification of the product in response to new assessments of the marketplace. This is irritating, but after all the goal of a company is to make money, even at the risk of inconveniencing its engineering staff. The second force stems from knowledge which is gained by implementing a portion of the design. Sometimes it is only by writing and testing some code that you realize that there is a better way to structure the program.

Both of these forces are difficult to resist. Clearly it is hard to say no to a hierarchy of managers who are acting, or believe they are acting, in the best interest of the company which after all issues regular paychecks. Equally hard to resist, although in a different way, is the internal force born of a personal desire to create the best possible design. Nevertheless, there are times when it is necessary to deny both forces for the sake of completing the project in a finite amount of time.

But when the time is right to change the design, the success of the project or even of the company may well hinge on the adaptability of the design. Since you can never tell when a design will be forced to change, it is vital from the outset to use flexible design techniques. All the techniques discussed in this book provide the necessary degree of flexibility for a design to survive from initial conception through the final stages of implementation.

It is not surprising that the design for a program could evolve while the program itself evolves, in the light of our day-to-day observation of nature. Natural systems such as bodies or ecosystems tend to be structured on many levels. Generally it is not the case that the levels emerge one from the other, with the higher levels appearing only after the lower levels are complete. Rather the various levels of a system co-evolve interactively. Let us look at an example. Our own existences, beginning at the atomic level, are organized on the levels of the molecule, the protein, the organelle, the cell, the tissue, the organ, and then the body. We observe that in both the evolution of the species and the embryonic development of the individual it is never the case that the lower levels strictly precede the higher levels once the cellular stage is passed. The levels evolve at the same time, with each level helping to create the conditions for the continued evolution of its neighboring levels and itself.

This is also the nature of software design. The original design creates the conditions for the development of the code, and the obseved behavior of the code provides feedback which may set in motion a change (evolution) of the design.

1.4 Why Study Design?

Since the reader has gone to the trouble of buying, borrowing, or stealing this book and reading so much of the first chapter, it would be inconsiderate of me to proceed without providing a personal motivation for studying real-time software design.

A first compelling reason is a matter of survival. Real-time programs have to be more and more complex as time goes on, and a thorough formal understanding of the art and science of software design may be the only way of coping with that complexity. Recent and continuing quantitative changes in hardware performance have produced, and will continue to produce, qualitative changes in the nature of real-time programs. In particular, as we have seen, complex programs have the potential of manifesting emergent behavior, which must be managed and controlled.

The second reason for studying this subject is much more positive: being able to design well enables you to produce a high-quality product. There are several components to high-quality in this context. A product must first and foremost conform to its specifications despite any obstacles encountered during the development cycle. Moreover, it should be completed in a timely manner. A product will always have personal value to its engineers, but in order to be of value to the corporate or scientific community it must be finished in time to be marketable and usable. (This is all the more important in a field where new introductions have such brief product lifetimes.) Lastly, a high-quality product must be maintainable. In the last section we saw how the specifications are subject to change after the development phase has begun; they are also subject to change after the development phase has ended. The original engineers may no longer be available when a change is required, so the product itself has to be learnable and adaptable. Organized systems are adaptable, and disorganized systems are not. Many programs have been as adaptable as dinosaurs, and have met a similar fate.

The last motivation for studying design is even more positive than the previous one. While it is true that engineers nowadays enjoy prestige, high salaries, and remarkable fringe benefits, these are all external rewards. What makes a job fulfilling is the sense of internal reward. Being able to design contributes to making engineering a personally rewarding experience. Designing software is good practice at general organizational skills, which increase your value to yourself as well as to your company and professional community. By designing your programs you can also accelerate your development as an engineer and an artisan, because you create the opportunity to learn from your design. In the last section we saw how hierarchies in a system co-evolve, each level helping the others to develop. Then we applied this principle to a two-level hierarchy with the program at the lower level and the design at the higher level. We can add a third level: the engineer. If you can develop from the interactions within the hierarchy, then you too can participate in the co-evolution of the various levels of your product.

Let me close this chapter by quoting Merlin in Mary Stewart's novel *The Crystal Cave*: "Magic is ninety-nine percent competent engineering."

CHAPTER 2
The Hardware Substrate

In this chapter we narrow our scope from the system-level speculations of Chapter 1 to the nuts-and-volts hardware level. This is more than a formality. It is impossible to write effective software for a real-time system without an understanding of the system's hardware, because the hardware and the software are too strongly related. The hardware scans all the relevant available states of the environment, encodes them, and reports them to the software; the software responds to the reports, part of the response being to output encoded instructions to the hardware. The job of the software is to decode the inputs into meaningful information, determine the appropriate reaction, and encode the reaction into output signals. All of this decoding and encoding is heavily dependent on the nature of the hardware with which the software is communicating.

The hardware devices which we will investigate in this chapter fall into three categories: processors, memory, and peripherals. We will not discuss fabrication techniques, performance specifications, or interfacing strategies. We are far more interested in logical properties than in electrical properties. While the software and the logical hardware levels are strongly coupled, the logical and the physical hardware levels are not. That's the whole idea of black-box integrated circuits. Besides, the differences are profound between similar devices (serial-to-parallel interfaces, for example) made by different manufacturers. In this book we will avoid detailed discussion of the products of any particular manufacturer.

The one exception to this rule will come shortly, when we discuss processors. In order to be able to give examples of assembly language code, we had to choose one processor. We selected the Intel 8086 microprocessor, and will briefly introduce it in the next section. This will by no means be a thorough introduction; its intent is to make the assembly language examples understandable to the reader who is unfamiliar with the 8Ø86.

2.1 Processors

A processor is a finite state machine. Its state changes according to a series of encoded instructions and data known as the program.

One of the processor's states, which changes every time an instruction is executed, is the address in memory of the next instruction to be executed. The result of an instruction could be a change in one of the processor's internal states, and/or an interaction with memory, and/or an interaction with input/output space.

How are the processor's internal states implemented? As registers. A register is a high-speed binary storage device, and in processors they are usually organized into groups of one, eight, or sixteen bits. Each bit is always in either the "Ø" state or the "1" state. Registers are depicted schematically as in Fig. 2.1.

Q is an output signal which reflects the state of the bit. D is a data input signal. When CLOCK makes a transition—typically from a high state to a low state—the state of the input at D is written into the memory bit. With an 8-bit register there are 8 D inputs, 8 memory bits, and 8 Q outputs, but only one clock signal. The same holds true for a 16-bit register.

Within a processor, registers communicate with one another by means of an internal network called a "bus." A bus is simply a set of interconnections among the registers. The idea of a bus is that in a single transaction or "micro-operation" the contents of one register (the "source") are driven from the register onto the bus, and from the bus are clocked into another register (the "destination"). An 8-bit processor has an 8-bit internal bus, a 16-bit processor has a 16-bit internal bus, and so on.

Actually, this is an oversimplification. A processor as described above could move data but could not operate on it: a computer which could not compute, a processor which could not process! Computation is provided by a device called an Arithmetic and Logical Unit, or A.L.U. Typical A.L.U.'s have two sets of data inputs, one set of data outputs, and a set of "select" inputs. The "select" lines dictate which operation the A.L.U. will perform (adding, logial OR'ing, decrementing, etc.) on its data inputs

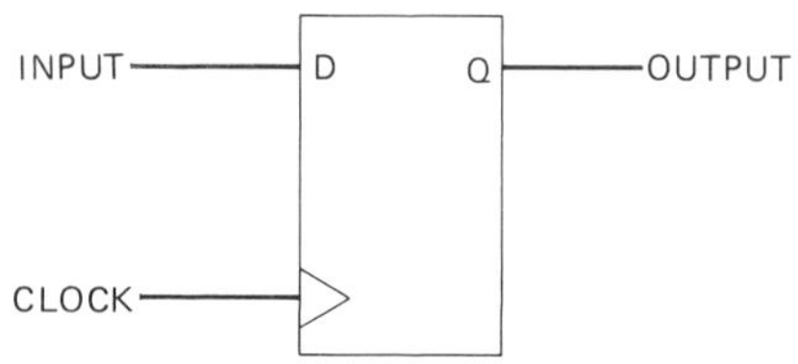

Figure 2.1

in order to generate its data outputs. The "select" inputs are determined by the instruction being executed.

An A.L.U. is depicted schematically in Fig. 2.2.

Within a processor, register outputs are driven to the A.L.U.'s data inputs, and the A.L.U.'s data outputs are connected to the register inputs. The select inputs are driven by a portion of the hardware called the *controller* or *sequencer*, which also is responsible for clocking data in and out of the registers. As you can guess, the actions of the controller are determined by the instruction being executed.

Figure 2.3 shows a schematic illustration of the connections between the registers and the A.L.U.

To visualize the movement of information in the above diagram, consider the Zilog Z-8Ø microprocessor. Two of its registers are named A and B. The instruction ADD A, B adds together the contents of these two registers and stores the result back in A. To execute this instruction, the controller would drive the contents of A to one input of the A.L.U., and the contents of B to the other input. It would select the ADD function of the A.L.U. After a brief interval to allow for propagation delay through the A.L.U., it would clock the result into the A register.

(A.L.U.'s include a function to pass one of the inputs straight through to the output regardless of the other input, so data can be moved from one register to another without being operated on.)

Three registers of particular interest are the Program Counter, the Memory Address Register, and the Memory Data Register. Remember that actual processors may not use the same names or implement things in exactly the same way.

The Program Counter (PC), to which we have already alluded, contains the address of the next instruction to be executed. Fetching the instruction is part of the responsibility of the Memory Address Register (MAR) and the Memory Data Register (MDR).

Just as the processor's internal bus connects the internal registers, there is an external system bus which connects together the processor, the

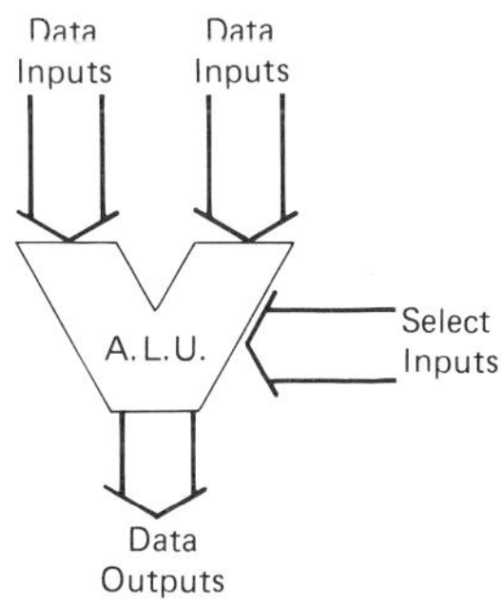

Figure 2.2

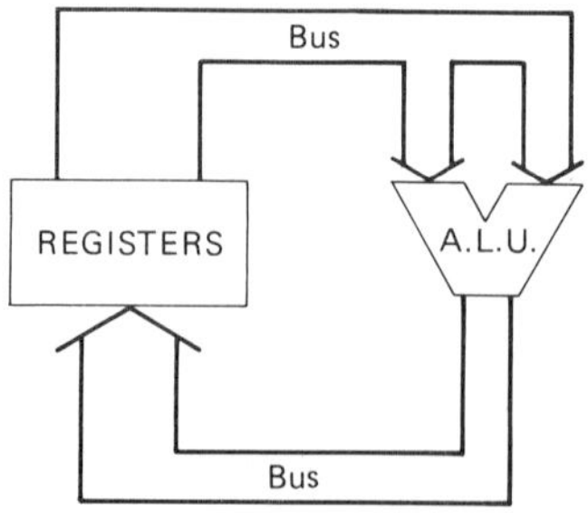

Figure 2.3

memory, and the input/output devices. The external bus carries not only data signals, but also address signals and control signals. System bus implementations vary widely.

The processor writes data to memory by driving the contents of the MAR onto the address lines of the system bus and the contents of the MDR onto the data lines, and asserting the proper control signals to implement the write operation. The processor reads data from memory by driving the contents of the MAR onto the address lines of the bus, and then asserting the propercontrol signals to implement a read; shortly the memory drives the contents of the addressed location onto the data lines, which can then be clocked into the MDR. (Driving the control lines, enabling the registers onto the bus, and clocking read data into the MDR are of course the task of the controller portion of the processor.)

Figure 2.4 shows a simple block diagram of what we have been discussing.

We have shown the PC as being connected to the MAR, and we have shown something called "DECODING" coming out of the MDR. (The PC, the MAR, and the MDR are also connected to the internal bus, but

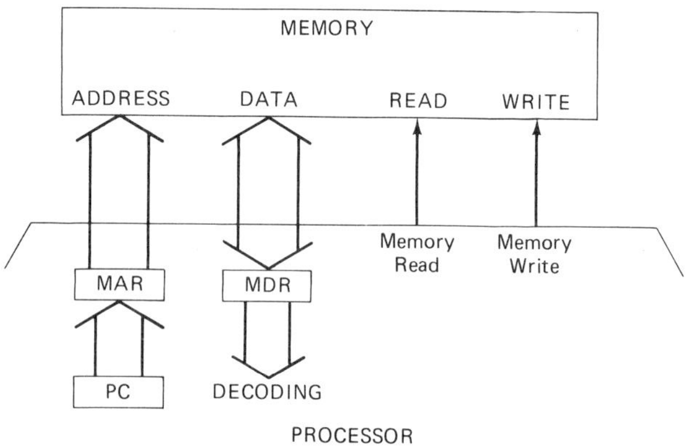

Figure 2.4

to draw this in would only confuse the picture.) When it is time to fetch the next instruction, the controller copies the contents of the PC into the MAR and then initiates a memory read operation. When the instruction is available from memory, it is clocked into the MDR; then it is decoded—broken down by the controller into the signals which will generate the proper set of micro-operations in the proper order so that the instruction will be executed correctly.

Early on in the decoding of the instruction, the PC is incremented so that it points to the next instruction. This incremented value might never be used; it may be overwritten by the controller. This is the case if the instruction being decoded turns out to be a jump, call or return instruction.

From a programmer's point of view, there are three features which distinguish one processor from another. These are the register set, the addressing modes, and the instruction set.

The register set is not the entirety of registers in the processor. Some of these registers, for example the Memory Address Register or the Memory Data Register, are strictly reserved for the processor's own use; the programmer does not have access to them, and does not need access to them. From a programming standpoint, the register set consists of those registers which are explicitly accessible via the processor's instructions. The number of registers varies considerably from processor to processor, as do the functions performed by the registers. In some processors all registers play the same role, while some processors assign a different function to each register.

The Zilog Z8 microcomputer, for example, has 144 eight-bit registers named "ØØ" through "7F" and "FØ" through "FF". Each of the first 128 registers is programmatically exactly like the others: whenever one of them may be used as an operand in an instruction, any of the others may legally be used in the same context. The Intel 8Ø8Ø microprocessor, on the other hand, has seven eight-bit registers named A, B, C, D, E, H, and L. Six of them can be combined pairwise—B with C, D with E, and H with L—to form sixteen-bit registers; these are referred to as BC, DE, and HL. There are certain instructions which can only operate on a subset of the register set. For instance, sixteen-bit addition may only be performed if HL is the destination register.

The second distinction among processors is the set of addressing modes. These are the various ways available to the processor for finding the operands of its instructions. Operands, which are simply the values on which an instruction operates, can be specified explicitly within the instruction, they can be found in registers, they can come from input/output devices, or they can reside in memory. The common types of addressing modes are called IMMEDIATE, REGISTER, I/O, ABSOLUTE, INDIRECT, and INDEXED. Each manufacturer has its own nuances for these terms.

IMMEDIATE ADDRESSING refers to the situation where the operand is specified in the instruction. For example, to add 5 to the contents of the CL register in the 8Ø86, we could use the instruction ADD CL,5. Here the 5 is specified by immediate addressing.

With REGISTER ADDRESSING the desired operand is found in a register. In the ADD CL,5 example above, the destination operand CL is specified by register addressing.

(The terms IMMEDIATE ADDRESSING and REGISTER ADDRESSING are actually a bit misleading. They imply that an address is being generated. In fact, no address is needed.)

I/O ADDRESSING refers to reading an operand from an input device or writing an operand to an output device.

All processors provide for some kind of immediate and register addressing; they would hardly be useful if they did not. I/O addressing is frequently provided; where it is not, the I/O devices in the system are connected so that they are indistinguishable from memory. (This scheme is called "memory-mapped I/O," and it relieves the processor of having to provide input and output instructions.) Where processors usually differ is in the modes they offer for accessing operands in memory. With ABSOLUTE ADDRESSING, the instruction simply specifies the address in memory where the operand resides. INDIRECT ADDRESSING specifies an address in memory whose contents in turn give the address of the operand. INDEXED ADDRESSING is like register direct addressing, but to the contents of the register may be added a constant or the contents of another register.

These addressing modes may be combined in many different ways. For example, we could concoct an INDEXED REGISTER INDIRECT mode, in which a constant is added to the contents of a register to produce an address in memory in which the address of the desired operand may be found.

To summarize the memory addressing modes, absolute addressing gets the address from the instruction, register direct addressing gets it from a register, indirect addressing gets it from memory, and indexed addressing has to calculate it.

The third distinction among processors is the instruction set, which is nothing more than the instructions which the processor can recognize and execute.

We have identified these three distinctions not for theoretical reasons but because when one is learning about a new processor with the intent of writing code for it, the three most important things to study are the register set, the addressing modes, and the instruction set. With this in mind, let us examine the processor we have selected for the coding examples in this book: the Intel 8Ø86.

The 8Ø86 has what is known as a pipelined architecture. This means that the processor consists of two independent parts: the Execution Unit and the Bus Interface Unit. The Bus Interface Unit manages an eight-byte queue on board the chip; at any given moment this queue contains the next several instructions from memory. The Execution Unit, which decodes and implements the instructions, gets bytes of code from the Bus Interface Unit, not directly from memory. Whenever the Execution Unit is not using the system bus for memory transactions, the Bus Interface Unit is reading bytes of program, trying to keep its queue as full as possible. This produces a considerable speed-up, because under ideal conditions (i.e., when the queue is nice and full) the Execution Unit only has to wait a miniscule amount of time for code; instruction fetching is almost instantaneous. The advantage is lost whenever a program branch occurs, because then the bytes in the queue are no longer valid and the queue must be emptied and refilled.

With this introduction to pipelined architecture in mind, let us dive into our investigation of the processor.

We will start with the register set. At first glance the registers seem to be something of a hodge-podge, because the 8086 is one of those processors in which each register serves a different purpose. There are thirteen 16-bit registers, with a fourteenth pseudoregister which is an assemblage of status bits. (Such a "flag" register is quite common among microprocessors.) The registers fall into four groups: four data registers, four pointer/index registers, the Instruction Pointer register (in a group by itself), and four segment registers.

The data registers are called AX, BX, CX, and DX. These are general-purpose accumulators, and may be used as operands in almost all arithmetic or logical instructions, as well as in data transfers to and from memory. The upper and lower bytes of each of them may be accessed independently as eight-bit registers. When used in this way, the low halves are called AL, BL, CL, and DL; the high halves are called AH, BH, CH, and DH. Only the data registers may be split up in this manner. All the other registers are accessible only as sixteen-bit wholes.

The pointer and index registers are also sixteen bit accumulators, in the sense that they are able to participate in arithmetic and logical operations as well as in data transfers with memory. They are particularly useful as pointers to operands in memory. (A "pointer" to a location in memory is simply a register or a memory word which contains the location's address.) The four pointer/index registers are the Stack Pointer, the Base Pointer, the Source Index, and the Destination Index.

The Stack Pointer (SP) always points to the top of the stack, which resides in memory and is a traditional last-in-first-out data structure. When a sixteen-bit value is to be "pushed" onto the stack, SP is decremented by 2 and the value is stored in the memory location now pointed to by SP. Note that the stack grows downward, from higher to lower addresses. When a sixteen-bit value is to be retrieved or "popped" off the top of the stack, the value is found at the location pointed to by SP; once that value is retrieved into the processor, SP is incremented by 2. The stack is a very useful instrument for temporary data storage. It is also the best scheme known for storing subroutine return addresses: when a call is implemented, the return address, which is the address of the instruction immediately after the call, is pushed onto the stack; to implement a return from the subroutine, the processor has only to pop the return address. Subroutine nesting and even recursion are possible with this scheme. Code generated by compilers usually uses the stack for passing subroutine parameters.

The Base Pointer (BP) can participate in a variety of addressing modes. It is especially useful for accessing memory data in the vicinity of the stack. We will see why this is the case when we look at the segment registers.

The Source Index register (SI) and the Destination Index register (DI) likewise participate in several different addressing modes. They are also used as pointers for several powerful string-processing instructions. One such string instruction two strings of contiguous bytes in memory; another copies a string to a different location.

The Instruction Pointer (IP) contains the address of the next instruction in the program to be executed. It is the 8Ø86's Program Counter.

The four segment registers are the Code Segment Register (CS), the Data Segment Register (DS), the Stack Segment Register (SS), and the Extra Segment Register (ES). They are needed to support an interesting feature of the 8Ø86: it has twenty address lines coming out of the chip, and is therefore capable of accessing an entire megabyte of memory. This is strange because all of the 8Ø86's registers are only sixteen bits wide. How can any useful addresses be generated? Where do the other four bits come from?

The segment registers provide the missing address information. A memory address is always generated by adding the address in the instruction to sixteen times a segment register. The particular segment register used depends on context. To take an example, whenever an instruction is to be fetched, the Bus Interface Unit generates its address automatically by multiplying CS by sixteen and adding IP. (From now on we will use the name of a register to denote its contents whenever this will not be ambiguous.) Whenever an instruction accesses memory, DS is multiplied by 16 and added to the address in the instruction. Whenever a value is to be pushed onto or popped from the stack, the relevant stack address is generated by multiplying SS by 16 and adding SP. ES is an extra segment register which can contain any value the programmer wants; it is useful for implementing a second data area.

The 64-Kbyte block of memory beginning at sixteen times a segment register's contents is called a segment. Thus we speak of the code segment, the data segment, the stack segment, and the extra segment. The value which is added to sixteen-times-the-segment-register is called the offset into the segment.

There is one entity in the 8Ø86 which might be called a "pseudoregister";

it is the FLAGS register. It is not a real register, but rather a composite of different status and control bits which are implemented as flip-flops scattered about the processor. They are connected to different lines of the internal bus so that for all practical purposes they seem to be the various bits of a sixteen-bit register. The only three of these flags which concern us are the Zero Flag (Z), the Carry Flag (C), and the Interrupt Enable Flag (I). Z and C are updated whenever the A.L.U. performs an operation. Z is set if the result of the operation was zero. C is set if there was a carry out of the most significant bit of the result, for example when two large numbers are added together. The I flag controls the interrupts. When I is set, the processor responds to interrupt requests from the external hardware; when I is clear, interrupts are disabled.

Having examined the 8Ø86's registers, we are in a good position to discuss its addressing modes. Register Addressing and Immediate Addressing are of course provided. The processor can also perform Direct Addressing and a wide assortment of Indexed Addressing modes. Indirect Addressing is not implemented.

The really interesting modes are the variety of Indexed Addressing modes. All you have to remember is that indexed addresses are generated by adding up an optional index register, plus an optional base register, plus an optional constant. The base register is either BX or BP (B stands for base). The index register is either SI or DI (I stands for index). The constant is sometimes an eight-bit value, sometimes sixteen, depending on the other parts of the address. If only a constant is used, with no base or index register, we have direct addressing.

You will recall that Intel uses the term "offset" to refer to the sum of the different components of an address, before the segment is taken into account. In general the segment refers to a location within the data segment—that is, the absolute location of the byte being referenced is sixteen times DS, plus the offset. There are, however, three exceptions. First, instructions which operate on two strings use SI to point to one string and DI to point to the other. The string pointed to by SI is in the data segment, but the one pointed to by DI is in the extra segment. Second, any address which is constructed using the BP register refers to a location in the stack segment. The third exception is a meta-exception: the programmer may override the normal segment assignments by specifying that one of the other three segments should be used instead.

We have summarized the various memory addressing combinations in Table 2.1. Whenever a displacement is specified, we have arbitrarily used a value of 99.

It is because of these segment registers that the 8Ø86 is so popular as a processor in multi-user microcomputer systems. A multi-user operating system has to assign memory dynamically as different users log on and off the system, running programs of various lengths and various RAM

Table 2.1

Mode	Offset	Segment	Symbolically
DISPLACEMENT ONLY	99	DS	.99
BASE ONLY	BX	DS	[BX]
	BP	SS	[BP]
INDEX ONLY	SI	DS	[SI]
	DI	DS	[DI]
DISPLACEMENT + BASE	99 + BX	DS	99[BX]
	99 + BP	SS	99[BP]
DISPLACEMENT + INDEX	99 + SI	DS	99[SI]
	99 + DI	DS	99[DI]
DISPLACEMENT + BASE + INDEX	99 + BX + SI	DS	99[BX + SI]
	99 + BX + DI	DS	99[BX + DI]
	99 + BP + SI	SS	99[BP + SI]
	99 + BP + DI	SS	99[BP + DI]

requirements. Suppose a user wants to run a program which accesses data stored at location 3ØØH. Suppose further that that location is already being used for data storage by a prior program. The operating system switches between one user and the next by storing all of the first user's registers, including the program counter, in a data structure reserved for the first user. Then it loads all the registers from the second user's data structure (they were stored there the last time the second user was suspended); it restores the second user's program counter last of all and voila! the processor is suddenly executing the second user's program.

Unfortunately this won't work in our example. The second program would alter the first program's data at 3ØØH, so the two programs must not be permitted to run concurrently. The solution to this general dilemma is to require that all user programs be RELOCATABLE. This means that all references to memory locations are made with respect to an arbitrary reference point whose actual value is unknown and immaterial. Thus a relocatable program would not store the contents of a register in memory location 7; instead, it would store the data in the seventh byte past the reference point, which would be assigned by the operating system. Thus the operating system can locate programs and their data in regions of memory where they will not be endangered by other programs.

For many processors it is extremely cumbersome to write relocatable code. A certain kind of programming discipline is needed to deliberately include the reference point in every memory access. Too much mental energy is expended on maintaining the discipline, energy which would be better spent on being creative.

This is just where the 8Ø86's segment registers come to the rescue. It requires no special techniques at all to write relocatable code for the 8Ø86. Any program is relocatable, provided it does not alter the CS, DS, or ES registers. These registers provide the reference point for relocating the program—or rather, three different reference points: one for the code, one for the data, and one for the stack.

Having examined the 8Ø86's register set and addressing modes, we are ready to look at its assembly language instruction set. Our intent is only to describe those instructions which are used in subsequent coding examples. The processor can execute 98 different instructions, and the best way to learn all of them is to read a book which is solely devoted to the 8Ø86. (Intel's User's Manual is the authoritative reference.)

The most basic instruction is MOV, which copies data into a memory location or into a register. Either a byte or a word may be MOVed. The value MOVed can be immediate (specified explicitly in the instruction), or it can come from a register, or it can come from memory. The only restriction is that data may not be MOVed from one location in memory to another. This is summarized in the table below:

		Source		
		Immediate	Register	Memory
DESTINATION	REGISTER	Allowed	Allowed	Allowed
	MEMORY	Allowed	Allowed	Not Allowed!

The syntax of the instruction in 8Ø86 assembly language (our standard here is ASM86 by Digital Research) is

```
MOV   destination,source
```

where "destination" is the register or memory location to receive the data, and "source" is either an immediate value or a register or memory address which contains the value to be MOVed.

If an address is expressed as the contents of a register, then the name is shown within square brackets, as we have seen in the addressing modes table. So for example the instruction to move into the AL register the byte in the data segment memory location pointed to by BX, would be written like this:

```
MOV   AL,[BX]
```

If the address is formed by adding a base and an index register, then both registers appear within the brackets, separated by an addition sign. Either register may come first. So the following instruction would copy the contents of DX into the memory word pointed to by SI plus BX:

```
MOV   [SI+BX],DX
```

If there is a displacement as well, then it appears before the open square bracket, as shown below:

```
MOV   7[BP+DX],AX
```

If moreover the specified address refers to a location in some segment other than the data segment, then the name of the "overriding" segment precedes the entire rest of the address, separated by a colon. The example below uses the most complicated addressing possible to specify the source operand:

```
MOV   AL,CS:45[SI+BX]
```

Nothing difficult about it. Just add up SI, BX, and 45, and use the result as an offset into the code segment. Copy the byte found there into the AL register.

Constants when used as operands may be expressed either numerically or symbolically. We will discuss symbolic constants a little later. Numeric constants are expressed explicitly as numbers. To move a value of 6 into the CH register, one would write:

```
MOV   AH,6
```

To use an explicit numeric value as an address (absolute addressing), precede the number with a period. This is necessary because the assembler needs a way to distinguish between immediate and absolute addresses. The following instruction moves into AH the contents of memory location 128 in the data segment:

```
MOV   AH,.128
```

The other arithmetic and logical instructions which we will find useful are ADD, SUB, AND, OR, XOR, CMP, TEST, INC, and DEC. Their operands are denoted in exactly the same way as those of the MOV instruction.

The first five of these instructions always operate on two operands, which are represented in the same "destination, source" notation as we user for MOV. They may operate on either bytes or words, provided the source and the destination are of the same type. CMP and TEST are a little bit tricky. They do not actually update the destination; they only change the status flags which reflect the latest A.L.U. operation (particularly C and Z). CMP updates the flags as if a SUB (subtraction) had been performed. TEST updates the flags as if an AND (boolean conjunction) had been performed.

INC and DEC take only one operand, which may be either a byte or a word. The contents of the specified register or memory location are incremented (INC) or decremented (DEC) by 1.

Table 2.2 summarizes the A.L.U. instructions which you will encounter in this book.

We will use three other kinds of instruction: Input/Output Instructions, Flag Instructions, and Branch Instructions. They are all much simpler than the A.L.U. instructions.

Input/Output instructions move data between peripheral I/O devices and AL or AX. The device's address in I/O space may be given either

Table 2.2

Opcode Mnemonic	Function	Number of Operands	Changes Destination
MOV	Move	2	Yes
ADD	Add	2	Yes
SUB	Subtract	2	Yes
AND	Logical AND	2	Yes
OR	Logical OR	2	Yes
XOR	Logical Exclusive OR	2	Yes
CMP	Subtract	2	No
TEST	Logical AND	2	No
INC	Increment	1	Yes
DEC	Decrement	1	Yes

numerically or symbolically. Symbols such as I/O device address may be defined using the "EQU" statement, which is a pseudo-op. Its syntax is

SSS EQU NNNNN

where SSS is the symbolic name and NNNNN is the number to be assigned to the name. Whenever the assembler sees the symbol SSS, it will use instead the number NNNNN.

The syntax of an input instruction is either

IN AL,address

or

IN AX,address

depending on whether the value to be input is a byte or a word. Similarly the syntax of an output instruction is either

OUT AL,address

or

OUT AX,address

Branch instructions alter the value of IP, thus causing the program to branch off from its normal linear order. Branching may be due to a subroutine call or return, or to a JUMP instruction. Jumps may be unconditional, or they may be conditional on the state of one or more of the flag registers. The conditions we will use in this book are the Carry flag being set or reset, and the Zero flag being set or reset. Subroutine calls and returns are always unconditional. All of these instructions, with the exception of the subroutine return, require that a destination address within the code segment be specified as an operand. Our subset of the branch instructions is summarized in Table 2.3.

Table 2.3

Opcode	Meaning	Conditional or Unconditional
JMP	Jump	Unconditional
JC	Jump if Carry Flag set	Conditional
JZ	Jump if Zero Flag set	Conditional
JNC	Jump if Carry not set	Conditional
JNZ	Jump if Zero not set	Conditional
CALL	Call subroutine	Unconditional
RET	Return from subroutine	Unconditional

The location to jump to is almost always specified as the label of an instruction. Labels are names of locations in the code segment. The name precedes the instruction, and is terminated by a colon. The following code decrements SI and, if the result is zero, moves a value of 1234 into DX; if the result of decrementing is not zero, then DX is incremented.

```
        DEC     SI       ; Decrement SI
        JZ      SIZERO   ; Jump if SI now zero
        INC     DX       ; SI not zero, increment DX
        JMP     MORE
SIZERO: MOV     DX,1234  ; SI = Ø, so DX ← 1234
MORE:   (more code goes here)
```

Note the traditional assembly language format of optional label, opcode, operand(s), and optional comments. Comments are preceded by a semi-colon.

What we need now to round out our examination of the 8Ø86 is an understanding of how data can be accessed in a useful way. We know how to read or write a word or byte in memory, provided we can construct its address with one of the addressing modes; but how do we specify the address of a desired variable, data record, or table?

It is the job of the assembler to allow symbolic representation of addresses. The 8Ø86 is not complicated or difficult in this respect, but its manner of representing symbols is unconventional. Readers who are used to traditional eight-bit assemblers may find it counter-intuitive.

We can approach the subject by asking the question, What is the data segment? On the hardware level we already know the answer: it is the 64-kilobyte space with respect to which most memory accesses are made. On the software level the answer is: it is where the variables live.

In general, an 8Ø86 program will include a portion of pseudo-op code called the DSEG. Pseudo-ops are assembly-language instructions which don't get translated into machine instructions; rather, they give the assembler information about how to assemble. The DSEG is a catalogue of names of variables. The DSEG also specifies the sizes of the variables; a variable may be a byte, a word, or any number of bytes or words. Suppose we want to write a program with five variables: a word called COUNTER, a byte called CHAR1, a byte called CHAR2, an 8Ø-byte string record called BUFFER, and a 128-word table called TABLE. The DSEG to define these variables to the assembler would look like this:

```
DSEG
COUNTER DW    1
CHAR1   DB    1
CHAR2   DB    1
BUFFER  DB    8Ø
TABLE   DW    128
```

The first line, which just says DSEG, tells the assembler that the DSEG follows. Each variable name appears as the first item in its own row. "DW" and "DB" mean, respectively, "define word(s)" and "define byte(s)". The number after the DW or DB is the number of bytes or words to allocate to the variable. The offset into the data segment of COUNTER is Ø, of CHAR1 is 2, of CHAR2 is 3, of BUFFER is 4, and of TABLE is 84. (Offsets are always in bytes.)

We can specify a variable's address by using the variable name as all or part of an address. The following instructions move the contents of CHAR1 into BUFFER without any mention of where in memory those variables reside:

```
MOV  AX,CHAR1

MOV  BUFFER,AX
```

We can specify the offset of a variable by preceding the variable name with the word OFFSET. The following code uses SI as a pointer to BUFFER and stores AL in the fifth byte of that record:

```
MOV  SI,OFFSET BUFFER

MOV  4[SI],AL
```

OFFSET can also be used to specify the offset of a label in the code segment, for example when reading a table or a string in the program.

Occasionally ambiguities arise as to whether an operand is to be treated as a byte or a word. For example, in the instruction INC [SI + BX], should the byte or the word beginning at SI + BX in the data segment be incremented? Any ambiguous pointer can be made into a pointer to a byte by preceding it with "BYTE PTR"; thus INC BYTE PTR [SI + BX] would increment the byte. Similarly, INC WORD PTR [SI + BX] would increment the word.

This ends our discussion of the 8Ø86. Having discussed processors in general and the 8Ø86 in particular, we will now look at some of the various peripheral hardware devices with which processors share board space. We will look at memory, counter/timers, and serial-to-parallel interfaces.

It is considerably easier to investigate memory than it is to investigate processors. All you can do with memory is read it and write it, and sometimes you can't even write it. Integrated-circuit memory devices fall into two categories—RAM, which can be read or written, and ROM, which can only be read. (Once for a prank one of the memory manufacturers published a spec sheet on a WOM device: a Write-Only Memory which could not be read.)

RAM is an acronym for Random Access Memory. The term Random Access refers to the property that any location in memory, selected at random, can be accessed (that is, read or written) in the same amount of time as any other location in memory, selected at random. This differentiates RAM from sequential-access memory devices such as disks or tapes. In sequential-access devices the data is stored and retrieved in a particular order, so that the time to access data depends on the data's physical location on the medium. For example, it takes longer to read a byte which is near the end of a rewound magnetic tape than it does to read a byte which is near the beginning.

All RAM chips have address input lines, bidirectional data input/output lines, and control input lines.

The control signals determine what the RAM is to do. In most devices these signals are high when the RAM is inactive, and are briefly pulsed low-high when the RAM is supposed to do something. The RAM responds to the low-going edge of the control signal. If the controls tell the RAM that it is being written to, then the RAM stores the data on the data lines in the location selected by the address lines. If the RAM is told that it is being read, then it drives the contents of the location selected by the address lines onto the data lines.

There are two schemes for implementing the control lines. The straightforward way has one signal called WRITE and one called READ; pulsing WRITE writes to the memory, pulsing READ reads from it. The other way is to have an input signal called STROBE (or some such name) and on called WRITE/READ. When STROBE is pulsed, the RAM either writes or reads depending on the state of the WRITE/READ control. Either way, two control lines are required.

On the hardware level, there are two families of RAM: static RAM and dynamic RAM. Static RAM operates just as we have described, and is the easier of the two families to design and program with. Dynamic RAM offers speed, cost, and density advantages over static RAM, but it has a quirk which makes it a bit more difficult to work with. In keeping with the tradition of anthropomorphizing whenever possible, the quirk is usually described in terms one would usually use to describe a person. We say that dynamic RAM tends to forget things, and must have its memory refreshed from time to time. In practical terms, this means that data stored in a dynamic RAM remains valid for only a few milliseconds unless an operation like a WRITE or a READ is performed. Any such operation will "refresh the memory" of the RAM, so that it will be valid for another few milliseconds. Any system with dynamic RAM must incorporate hardware or software to take care of the refresh operation.

In general there are three uses for RAM: program storage, data storage, and nonvolatile data storage.

General-purpose computer systems store programs on disks; when a program is to be executed, it is copied from disk to RAM, where it is executed. Real-time systems usually do not employ this technique, because there is usually only one program to run; that is why most real-time programs execute out of ROM, which we will discuss shortly. Executing out of RAM has the dubious advantage of permitting self-modifying code, a risky technique by which an instruction is overwritten with another instruction; for example, an ADD opcode might be changed to a SUBTRACT opcode.

The most common use for RAM in real-time systems is storage of variables and data structures, Since these are subject to change, they must of course reside in memory which can be altered.

Modern real-time systems usually can be configured by the user into various modes of operation. With an alphanumeric color terminal, for example, the user can generally select the foreground color and the background color, as well as a number of communications parameters such as baud rate, parity, and protocol. Informing the system of the desired choices is laborious, especially if there are many choices to be made, and it is desirable to relieve the user of having to configure the machine every time it is turned on. The alternative is to have a small RAM in which the most recent configuration choices are stored symbolically. This RAM is powered all the time: when the system is turned off, a small battery provides power to the RAM. When the system is turned on again, the con figuration is read out of the RAM, which always contains valid data. Designers usually choose CMOS RAM for this purpose because the low power requirement of CMOS provides the maximum possible lifetime for the battery. (A cheaper and more reliable alternative to the battery-and -RAM implementation has recently appeared. We will shortly discuss the Electrically Eraseable Programmable Read-Only Memory, or EEPROM.)

Read-Only Memory, or ROM, is memory which can only be read from. The only control signal needed is a READ line. When this line is pulsed, the contents of the location selected by the address lines are driven onto the data lines. The density of readily available ROM I.C.'s has pretty nearly doubled every year since about 1977.

Something must be wrong here. Something is missing. If ROM can NEVER be written to, how can any useful data be stored in it?

The answer to this question depends on the kind of ROM being referred to. There are four subspecies of ROM: ordinary ROM, PROM, EPROM, and EEPROM.

Ordinary ROM is programmed by the vendor at the factory. The buyer supplies a pattern of 1's and Ø's which completely specifies the desired state of each bit in the memory. The manufacturer creates an I.C. mask for the pattern and mass-produces the ROM's, which are essentially customized integrated circuits. The charge for this service, above and beyond the price of the ROM's themselves, is called the "mask charge"; in the United States, this can amount to at least several thousand dollars. If a program with a bug in it is mass-produced, the mask charge is wasted. When this happens in a corporate environment, fingers are pointed and managers are spoken to sharply by higher managers. This is just human nature. Thus managers are rightly nervous about committing a program to ROM before it has been thoroughly checked out.

PROM stands for PROGRAMMABLE ROM. PROM's have high-voltage special control inputs which are similar to the WRITE inputs of RAM's. If a sufficiently high signal appears at such an input for a long enough interval of time, the data on the PROM's data lines (which are ordinarily strictly outputs) is written to the location selected by the address lines. In practice, ROM's are programmed by inserting them into machines called PROM Programmers. The pattern of 1's and Ø's is entered into the PROM Programmer, which performs the operation of programming the pattern into the PROM. In an unprogrammed PROM, fresh off the shelf, all the bit's are 1's; when the PROM is being programmed, only bits which are supposed to be Ø's are changed. Changing a bit from a 1 to a Ø is irrevocable. If a PROM contains an erroneous Ø, it can only be thrown away.

The difference between PROM's and EPROM's is that EPROM's can be erased and reused again and again. The "E" stands for Eraseable. EPROM's are very easy to recognize: in the center of the I.C. package is a small round clear plastic window. Through the window you can see the chip itself. The EPROM is erased by exposing the chip to high-intensity ultraviolet radiation, which catalyzes the process of changing all the bits in the memory to 1's. EPROM's can be exposed to ordinary indoor artificial light without losing data, and they can stand a few minutes of sunlight, but don't take them sunbathing all day in Puerto Vallarta.

EPROM's are erased with special lights, with chambers for the chips and interlocks to protect the user's eyes, and the process may take from a few minutes to an hour. An alternative method of erasure is provided by the EEPROM (pronounced with a glottal stop between the E's so that it will sound different from EPROM). EEPROM's are Electrically Eraseable PROM's, and as the name implies they are erased electrically rather than optically. EEPROM's can be used in any application where EPROM's are used. Moreover, if the circuit where the EEPROM resides is equipped with the proper erasure circuitry, then the EEPROM can be used for non-volatile long-term data storage, in the place of the RAM-plus-battery scheme we discussed earlier.

The most common use for Read-Only Memory in real-time systems is of course program storage. This is consistent with our observation that, since real-time systems tend to have only one program to run, they do not need to load code into RAM before executing it.

We will now round out our survey of the hardware substrate with a look at two of the more common kinds of peripheral I/O devices found in real-time systems. We will not be too specific, since implementations by different manufacturers vary so much. Often a manufacturer will create a line of peripherals which are designed to interface easily with their microprocessor. Thus for instance Zilog produces a Counter-Timer Circuit (the Z8Ø-CTC) and a Serial Input/Output chip (Z8Ø-SIO), both of which communicate in a straightforward way with the Zilog Z-8Ø processor.

The purpose of a peripheral device is to inform the processor of what is going on in some dimension of the external world (input) and to synchronize and coordinate the processor's commands to the environment (output). Typically there are two kinds of input information which a processor can read from a peripheral device: status and data. "Status" refers to the state of the device, and data is the actual data which it is the device's job to deliver to the processor. So for example a Floppy-Disk Controller I.C. could report status on whether the disk-drive door is open or closed or whether the drive motor is spinning or not, in addition to reporting the actual data on the disk. Similarly, the processor can perform two kinds of output to the peripheral: commands and data. Commands tell the peripheral how to operate, and data is of course the data to be delivered to the outside world.

A processor communicates with its peripherals in two different phases. The first phase is initialization, which occurs at the beginning of the system's program, right when the system is turned on. During intialization, the processor writes a number of commands to the peripheral, telling it what its operating modes shall be, so that it will be all ready to function. The second phase, which lasts until the system is turned off or restarted, is the device's steady-state operation.

A very efficient way for a peripheral device to pass information to the processor is by interrupting. We assume that the reader is already familiar with interrupts. Peripherals usually initiate interrupts under three sets of circumstances: when data in available for input to the processor, when the device is ready for output from the processor (having completed the previous output operation), and when the device's status changes.

With this introduction in mind, let us examine Counter/Timers and Parallel-to-Serial Interfaces.

Counter/Timers are glorified down-counters. Recall that a down-counter is a register composed of some number of bits. A LOAD signal writes these bits with a value specified on the counter's data input lines, and a COUNT signal decrements the value in the register by 1.

Counter/Timers enlarge on the properties of down-counters in three ways: they provide various ways to load the initial value, they allow for different methods of clocking the counter down, and they take various sorts of action when the count in the register reaches zero.

The first time after initialization that the Counter/Timer is used, the starting value of the register must be written by the processor. Most Counter/Timers remember their most recent starting value, and subsequent operations can optionally specify that this value should be used again. Thus as long as the starting value does not change, it need only be written once to the Counter/Timer.

There are typically two ways to clock the counter. Either the system clock (the signal which clocks the processor) can be used, or some other signal from elsewhere in the system can be used. The first method is called Timer Mode, the second is called Counter Mode.

What should a Counter/Timer do when it reaches zero? It should inform some entity somewhere in the system. It can either tell the processor, or tell something else. The way to tell the processor is to interrupt it. The way to tell the rest of the system is to pulse an output signal. Usually Counter/Timers can be configured to interrupt the processor, to pulse the output signal, or both or neither. Having spread the word that it has reached zero, the Counter/Timer may or may not start counting again from its old initial value, depending again on how it has been configured.

Are Counter/Timers actually useful? Indeed they are. They put the "Time" into "Real Time." They are widely used in three areas: as signal generators, as alarm clocks, and as real-time clocks. We look at all three uses.

Systems often need a pulse-train whose period depends on circumstance or on the user's selection. In other words, the period is not absolute. We will see that many devices which communicate serially with their environment offer the user a choice among several communication speeds or "baud rates." The baud rate is a function of the frequency of a clocking signal which is fed to the system's Serial I/O controller. This is a natural application for a Counter/Timer: whenever the processor determines that a new baud rate is required, it configures the Counter/Timer to run continuously, generating output pules but not interrupting, with the appropriate initial value dictating the appropriate period of the clocking signal. (If you are unfamiliar with serial communications and this example has been confusing, don't worry. We will cover serial communications at the end of this chapter.)

The Counter/Timer can also be used as a sort of "alarm clock." As an example of this application, consider a program which needs to assert an output signal for, let us say, 5Ø milliseconds. One way to do this would be to turn the signal on, busy-wait for 5Ø milliseconds, and then turn the signal off. With this strategy the processor can do nothing useful except count down for rather a long time. However, if the system has a Counter/Timer, the processor can devote a minimum amout of effort to the timing operation. It turns on the signal, and then configures the Counter/Timer to count down from some appropriate initial value so that in 5Ø milliseconds it will interrupt and stop counting. The program's interrupt routine turns off the output signal.

If there are many operations which need to be timed, a single Counter/

Timer will still suffice if it is treated like the Real-Time Clocks found in most computer operating systems. Suppose a system has a bell, an L.E.D., and a motor, any or all of which may need to be turned on for some given duration of time. Suppose the system has a single Counter/Timer, which is being clocked at a frequency of 1KHz. Suppose, lastly, that there are three output devices named BELL, LED, and MOTOR; conveniently, writing "1" to any of these outputs turns on the corresponding device, and writing "Ø" turns it off.

Let the Counter/Timer be configured to run continuously, interrupting whenever it reaches zero, with an initial value of 1Ø. Thus the processor will get an interrupt every hundredth of a second. How can we use these interrupts to time the bell, L.E.D., and motor operations?

One way, which is quite efficient, requires four variables: three counters (one for each device) and a flag byte. This flag byte, which we will call TFLAGS, is initially zero. If bit 7 is set, it means that the bell is on and is being timed; if bit 6 is set, the L.E.D. is on and is being timed; if bit 5 is set, the motor is on and is being timed. The other bits are always Ø.

To turn on one of the output devices for, say, N seconds, the software first writes a "1" to the appropriate output address to physically turn on the device. Then it writes N*1ØØ to the appropriate counter. (Let's call the counters BELLCTR, LEDCTR, and MOTORCTR.) Lastly it sets the appropriate bit in TFLAGS. The interrupt routine checks each bit in TFLAGS. If any bit is set, the corresponding counter is decremented; if the counter decrements to zero, the corresponding output device is turned off and the bit in TFLAGS is cleared. In a sense, the software is using one Counter/Timer to simulate the operation of three of them.

Let's look at the flowcharts in Figs. 2.5—2.8 and the accompanying code to accomplish this:

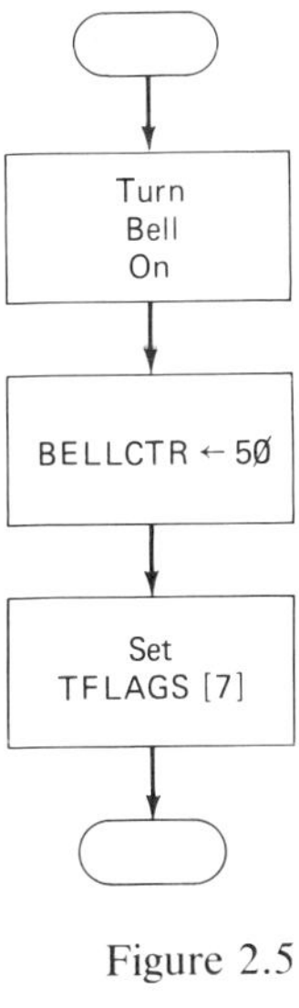


```
MOV    AL, 1
OUT    BELL, AL      ; Turn Bell on
MOV    AL, 5Ø
MOV    BELLCTR, AL   Counter ← 5Ø
MOV    AL, 8ØH
OR     TFLAGS, AL    ; Set TFLAGS [7]
```

Figure 2.5 Turning the bell on for ½ second.

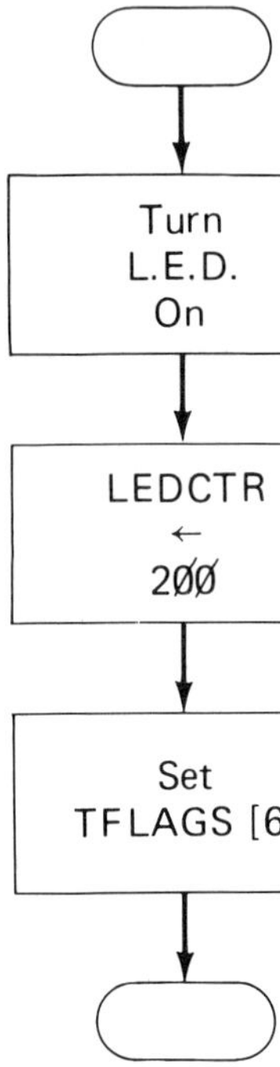


```
MOV     AL, 1
OUT     LED, AL      ; Turn L.E.D. On
MOV     AL, 2ØØ
MOV     LEDCTR, AL   ; Counter ← 2ØØ
MOV     AL, 4ØH
OR      TFLAGS, AL   ; Set TFLAGS [6]
```

Figure 2.6 Turning the L.E.D. on for 2 seconds.

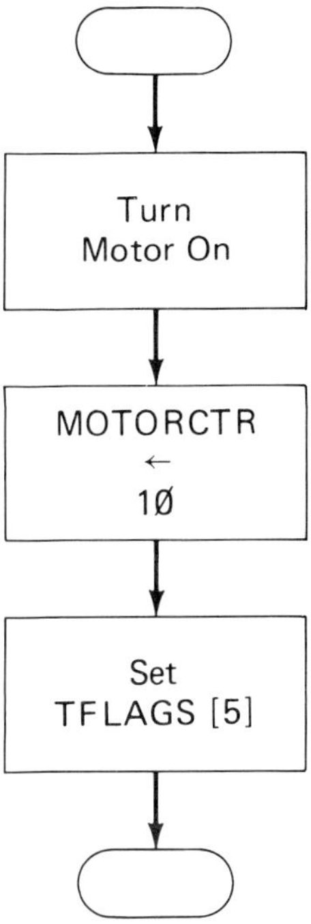


```
MOV     AL, 1
OUT     MOTOR, AL         ; Turn Motor On
MOV     AL, 1Ø
MOV     MOTORCTR, AL      ; Counter ← 1Ø
MOV     AL, 2ØH
OR      TFLAGS, AL        ; Set TFLAGS [5]
```

Figure 2.7 Turning the motor on for 1/10 second.

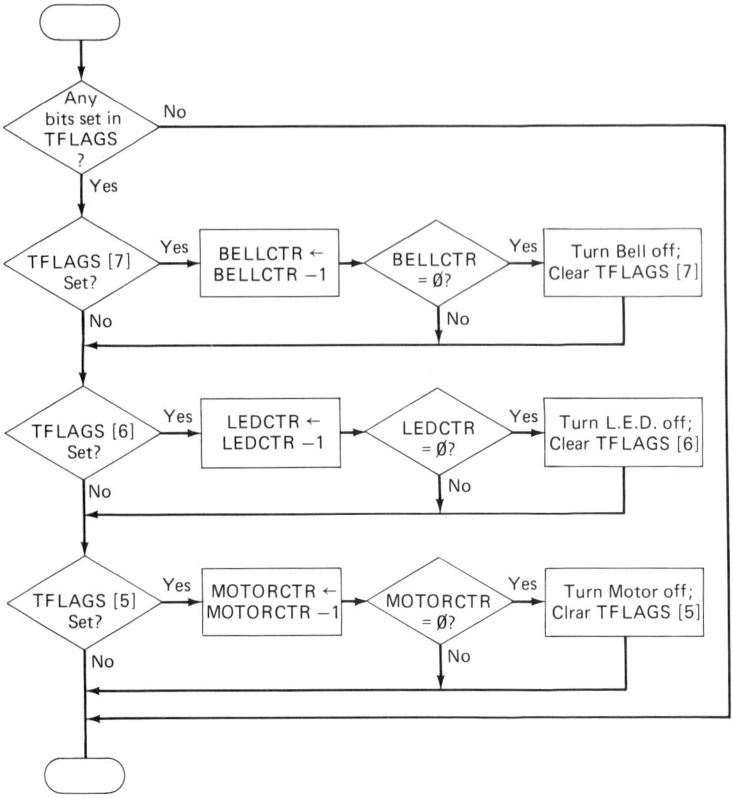

Figure 2.8 Counter/timer interrupt handler.

Take a look at the second and third lines of the interrupt routine (OR AH,AH and JZ TIMER_ INT_ Z). They correspond to the first decision diamond of the flowchart ("Any bits set in TFLAGS?"). At first glance they are superfluous: the routine would do exactly the same thing if those lines were eliminated. In fact, the lines are LOGICALLY superfluous; they are included to improve the system's performance. Remember that the interrupt happens every 1Ø milliseconds, whether or not any of the devices is on. Every microsecond required by the interrupt routine steals one-hundredth of one percent of the processor from the system (one microsecond out of every ten milliseconds). This doesn't sound like much, but it adds up. In a high-performance system, any routine which is executed periodically, no matter how infrequently, is a candidate for optimization. In our example, the second and third lines provide for a quick exit from the interrupt routine in case there is nothing to do.

It is quite common in real-time programs for interrupt routines to interact with background (noninterrupt) routines. In our example the background code consists of the three short routines to turn on the bell, L.E.D.

```
TIMER_INT:
        PUSH AX                         ; SAVE AX REGISTER
        MOV  AH,TFLAGS                  ; AH ← TIMER FLAGS
        OR   AH,AH                      ; ANY BITS SET IN TFLAGS?
        JZ   TIMER_INT_Z                ; NO
        MOV  AL,Ø                       ; YES
        TEST AH,8ØH                     ; BELL RUNNING?
        JZ   TIMER_INT_A                ; NO
        DEC  BYTE PTR BELLCTR           ; YES, DECREMENT BELL COUNTER
        JZ   TIMER_INT_A
        OUT  BELL,AL                    ; IF COUNTER=Ø, THEN TURN OFF BELL
        AND  AH,7FH                     ; AND CLEAR BIT 7 OF TFLAGS
TIMER_INT_ A:
        TEST AH,4ØH                     ; L.E.D. RUNNING?
        JZ   TIMER_INT_B                ; NO
        DEC  BYTE PTR BELLCTR           ; YES, DECREMENT L.E.D. COUNTER
        JZ   TIMER_INT_B
        OUT  LED,AL                     ; IF COUNTER=Ø, TURN OFF L.E.D.
        AND  AL,ØBFH                    ; AND CLEAR BIT 6 OF TFLAGS
TIMER_INT_ B:
        TEST AH,2ØH                     ; MOTOR RUNNING?
        JZ   TIMER_INT_C                ; NO
        DEC  BYTE PTR MOTORCTR          ; YES, DECREMENT MOTOR COUNTER
        JZ   TIMER_INT_C
        OUT  MOTOR,AL                   ; IF COUNTER=Ø, TURN OFF MOTOR
        AND  AH,ØDFH                    ; AND CLEAR BIT 5 OF TFLAGS
TIMER_INT_ C:
        MOV  TFLAGS,AH                  ; UPDATE FLAGS
TIMER_INT_ Z:
        POP  AX                         ; RESTORE AX
        STI                             ; ENABLE INTERRUPTS
        IRET                            ; AND RETURN FROM INTERRUPT
```

and motor, and this code interacts with the interrupt routine. Both "levels"—interrupt and background—access the same variables. It is vital, when developing such code, that the background and interrupt routines should function correctly even when the interrupt routine interrupts the very background routine with which it interacts.

You can verify for yourself that, in our example, if the interrupt routine is executed between any two consecutive lines of the background code, everything will work satisfactorily. But suppose we want to turn the motor on for only one hundredth of a second. If we change the order of events on the background routine as shown below, we could have a disaster on our hands:

```
MOV   AL,2ØH
OR    TFLAGS,AL        ; SET BIT 5 IN TFLAGS
MOV   AL,1
MOV   MOTORCTR,AL      ; MOTOR COUNTER ← 1
OUT   MOTOR,AL         ; TURN ON MOTOR
```

What happens if the interrupt occurs while the fourth instruction (MOV MOTORCTR,AL) is executing? Then the interrupt routine will be executed after MOV MOTORCTR,AL and before OUT MOTOR,AL. The interrupt routine will see that bit 5 of TFLAGS is set, so it will decrement MOTORCTR from 1 to Ø and turn off the motor. The motor is already off, so it will stay off. On returning from the interrupt, the next instruction will be OUT MOTOR,AL which will turn the motor on again. Nothing will ever turn it off again. We wanted it on for ten milliseconds, and instead it will stay on forever!

We end this chapter with a look at Parallel-to-Serial Interfaces.

Imagine a system which communicates data in eight-bit units to another device, along a cable. There are two possible implementations of interest. First, the cable could have eight data lines and one clock line. (For practical reasons, any such cable would also need at least one ground line, but we will ignore such considerations in this discussion.) To write a byte of data to the remote device, our system drives the data onto the cable's data lines and then pulses the clock signal to inform the remote device that the data lines are valid and should be read. Such an implementation might look like Fig. 2.9.

The remote device uses the CLOCK signal to write the data into an 8-bit register.

If the remote device is at any appreciable distance from our system, the cost of such a cable becomes prohibitive. Cable lines are expensive, particularly if the two systems are communicating over a telephone network. Moreover, each line requires a fairly sophisticated driver and receiver, and these drivers and receivers consume a lot of power.

In view of these considerations, most systems communicate serially: all bytes to be transmitted are sent one bit at a time, in a stream, along

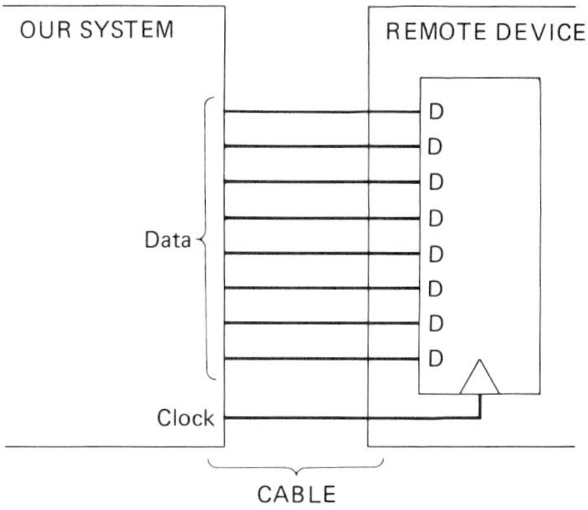

Figure 2.9

a single wire. The devices which convert the bytes to streams of bits and back again are called Parallel-to-Serial Interfaces. We conclude this chapter with a discussion of Parallel-to-Serial Interfaces and their uses in a synchronous communications.

Figure 2.10 shows two microprocessor-based systems communicating serially. Notice how much smaller the cable is. (We have again left out the ground signal.)

Parallel-to-Serial Interfaces consist of two sections: a transmitter, which converts parallel data to serial, and a receiver, which converts serial data to parallel. Both sections are essentially shift registers. When a byte is transmitted, the sending Interface reads it in parallel form, converts it to a serial bit stream, and transmits it along the cable. The Interface on the receiving end converts it back to parallel form and reports it to its processor. Both shift registers must be clocked at the same frequency. This is most often accomplished by human intervention, when the users configure the systems. The frequency at which the shift registers are clocked is called the "bit rate" or the "baud rate." One period of this clock signal is called one "bit time."

The transmitter section begins transmitting by loading the byte to be sent into its shift register. Then it shifts the byte, bit by bit, onto its serial output line, starting with the least significant bit. Each successive bit remains on the line for one bit time. Figure 2.11 shows a byte of 10011111 being transmitted.

Similarly, the receiver section receives data in its serial input; when all eight bits have been shifted into the shift register, the processor in the receiving system can inspect them as a byte by reading them in parallel form out of the shift register. Figure 2.12 below shows the byte 10011111 being received.

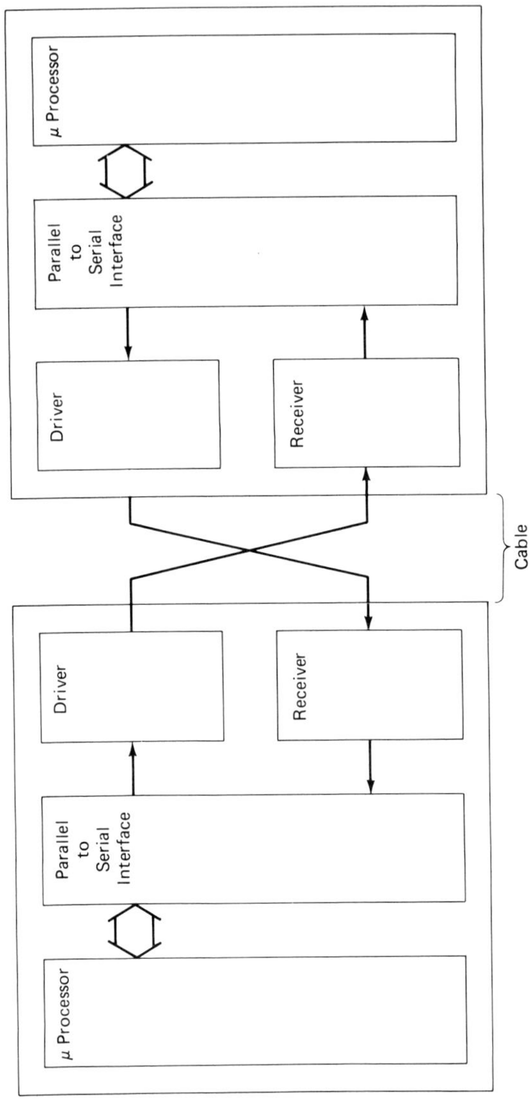
μ Processor
Parallel to Serial Interface
Driver
Receiver
Cable
Driver
Receiver
Parallel to Serial Interface
μ Processor

Figure 2.10

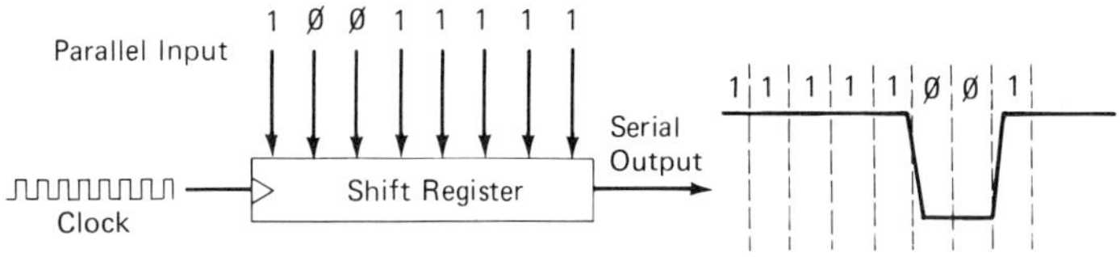

Figure 2.11

When the transmitter section is idle—that is, when it is not transmitting anything—it outputs a "1."

At this point you may have realized that this can't be the entire picture. Something must be missing from our description of the transmitter, and something else from our description of the receiver. Since the transmitter outputs a constant high level when it is idle, what is the difference between sending a byte consisting of eight 1's and just being idle for a while? And as for the receiver, how does it know when to start clocking in data? What prevents it from erroneously clocking in the 1's which the transmitter sends when it is idle?

Both of these problems are solved with an invention (which has since become a convention) called a "start bit." The start bit is the essence of asynchronous communications. (The prefix "a" in "asynchronous" means "not." Asynchronous communications are not synchronous in the sense that the receiving device never knows when the transmitting device might send something.) Before sending a byte, the transmitter drives the serial output line low for one bit time. This low level is called the start bit. It is a bit only in the sense that it is a level which stays on the line for one bit time. The start bit tells the receiver that more bits are coming. The receiver waits 1½ bit times from the leading edge of the start bit, and then begins clocking in data bits once every bit time. (The receiver waits one bit time so that the start bit will be finished, and then an additional half of a bit time so that it can clock in the data bits right in the middle of their lifetime, far from any edges.)

After the receiver has received an entire byte, it needs a little time to recover. To provide for this, the transmitter must remain idle for either

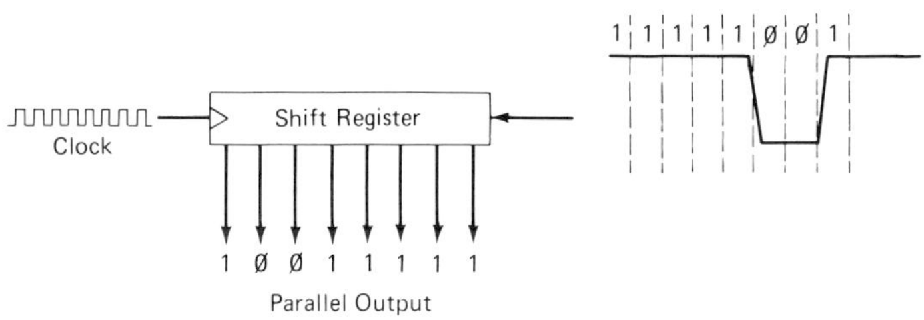

Figure 2.12

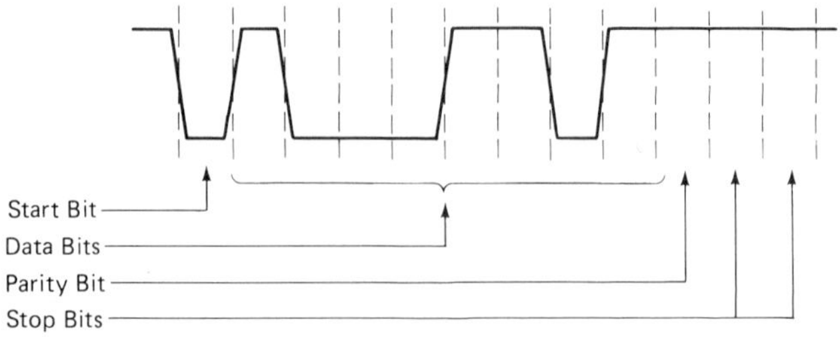

Figure 2.13

one, one-and-a-half, or two bit times. Each of these idle states is called a "stop bit," so we refer to the Parallel-to-Serial Interface as being configured for one, one-and-a-half, or two stop bits (even though the idea of half a bit doesn't make much sense.) Transmitters must be configured to send at least as many stop bits as are required by the receiver.

In addition to the start bit and the stop bit or bits, there is one more bit which may be transmitted. This is the parity bit, which if present is sent after the most significant data bit and before the first stop bit. The state of the parity bit is a function of the data bits, excluding the start and stop bits. There are two types of parity. With "odd parity" (which is the more common) the parity bit is 1 if there are an even number of 1's one in the data byte; the parity bit is Ø if there are an odd number of ones in the data byte. Even parity is just the opposite. Thus with odd parity an odd number of 1's is always sent, and with even parity an even number of 1's is always sent. The parity bit provides some protection against misunderstanding due to distortion to the serial signal while it is en route. If one bit gets changed, the parity bit will no longer be valid; the receiver can report the situation to the processor, which can reject the damaged byte or request the sending system to transmit it again. If parity is used, both Parallel-to-Serial Interfaces must be configured for the same kind of parity.

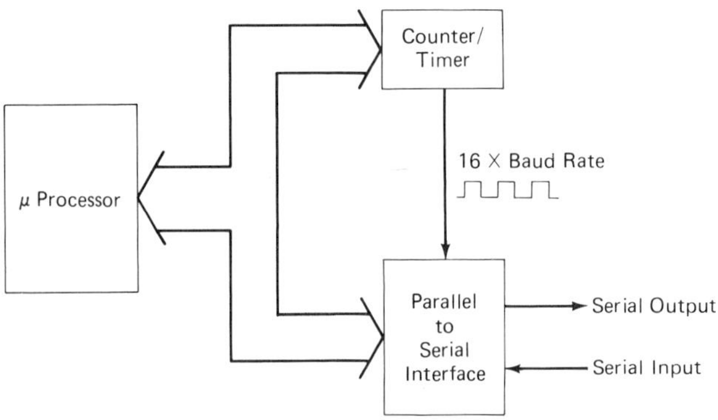

Figure 2.14

Figure 2.13 shows the complete waveform, including start, stop, and parity bits, which would be generated by a transmitter sending a byte equal to 1Ø11ØØØ1, with odd parity and two stop bits. Since parity is odd and there are four bits set to 1 in the data byte, the parity bit is 1. The time divisions are bit times.

One last aspect of Parallel-to-Serial Interfaces is clocking. We have mentioned that when the receiver detects the leading edge of a start bit, it waits 1½ bit times before beginning to sample data. Also, some transmitters have the capability of sending 1½ stop bits. Clearly, with all these halves of bit times, both the transmitter and the receiver must have access to a clock signal whose period is a fraction of a bit time. In practice, most Parallel-to-Serial Interfaces are driven by a signal which is 16 times faster than the desired baud rate. The receiver waits 24 clock periods after detecting a start bit, and subsequently samples data every 16 periods.

The clock signal is most often generated by a Counter/Timer, as shown in Fig. 2.14.

CHAPTER 3

Design Languages

In the last chapter we approached our subject from below, so to speak, with a look at the hardware substrate of real-time systems. We will now approach it from above by examining several high-level design languages.

As we mentioned in Chapter 1, annotated pictures are a more effective medium for software design than words alone. In the spirit of this observation, the four design languages we will discuss are all diagrammatic. We will look at Flowcharts, Structure Charts, State Diagrams, and Data Flow Diagrams.

Before we begin, it is important to understand the difference between a software design language and a computer language. A computer language is formal. Its purpose is to encode algorithms. It has a rigidly defined grammar and syntax, the slightest deviation from which—even a spelling mistake—generates an error. A design language, on the other hand, is informal. Its purpose is to encode not algorithms but ideas. Moreover, the design is read and searched for meaning not by computers but by people. Thus it must be more flexible and more subtle.

In this chapter we describe some design languages rather than attempting to define them. If in the course of designing a program you find it useful to alter or add to a design language, it is important to feel free to do so.

3.1 Flowcharts

The flowchart is the oldest and most common diagrammatic design language. We have already presented several flowcharts, since we assume that the reader is at least slightly familiar with this language.

A flowchart represents the flow of execution of instructions in a program

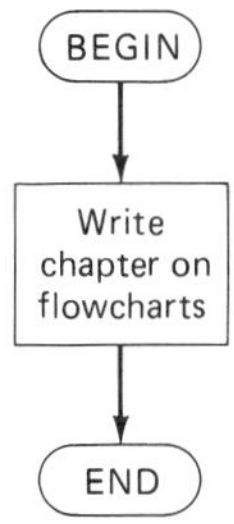

Figure 3.1

(this is known as the "flow of control"). Where a program branches in two directions, the direction actually taken during execution being determined computationally, the flowchart also splits in two. Thus a flowchart shows the order of execution of the various parts of the program in all possible situations.

The most common flowchart symbol is the *process box*. A process box represents a routine, a subroutine, or any contiguous set of instructions which are conceived of as a unit. A process box representing a specific bit of code is simply a rectangle with text inside it to describe what the code does. For example, if I wanted to write a program on how to write this section of this chapter (including this parenthetical remark) I would flowchart it as in Fig. 3.1.

A flowchart always begins with a bubble. The bubble can be empty, it can contain the name of the routine being represented, or as in our example it can contain some word such as BEGIN or START. Flowcharts may end in several places, since code may branch off; wherever the flow ends there should also be a bubble. An ending bubble may be empty or it may contain some appropriate legend such as END, DONE, STOP, or EXIT.

Order of processing is represented by arrows. A box may have only one arrow coming out of it, and after the box is executed the symbol at the head of the arrow is executed. (Using the metaphor of passing control, we could say that control passes to the symbol at the other end of the arrow.) There may be many arrows leading into a symbol; there must be at least one. It is good technique to have arrows point downward as much as possible, so that the flowchart can be read from top to bottom.

The flowchart to write this entire chapter would look like Fig. 3.2.

A computational or logical question whose outcome determines whether or not the program will branch off is represented by a diamond-shaped box. The question itself is written inside the diamond. Two arrows leave the diamond; one is labelled "YES" and the other is labelled "NO." The "YES" arrow points to the portion of the flowchart to be executed if the answer to the question is YES, and similarly the "NO" arrow points to the portion to be executed if the answer is NO. (If the question cannot

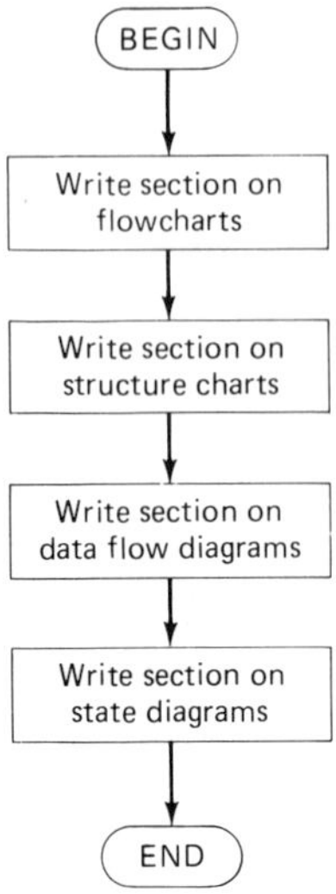

Figure 3.2

be expressed in a binary yes/no form, then it cannot be resolved by a computer and there is no need to try to flowchart it. A complicated question is only computable if it can be broken down into a finite number of binary questions. Multiple conditional tests and jumps, such as those provided by FORTRAN and PASCAL, get compiled into groups of binary tests.)

Using decision diamonds, the flowchart for the program to write this chapter (including this parenthetical question?) could be redrawn as shown in Fig. 3.3.

These are the main flowchart symbols, and with them you can represent

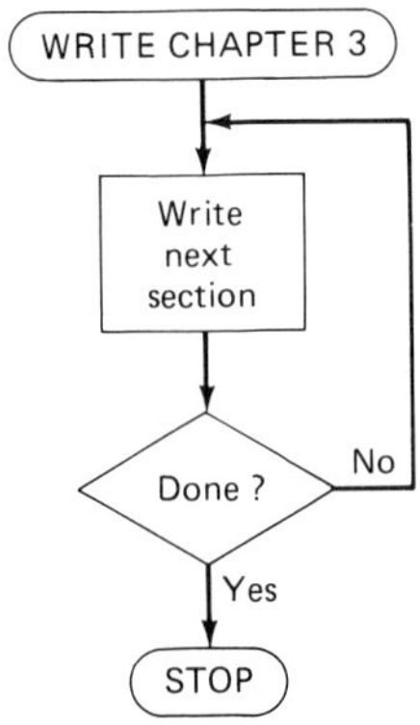

Figure 3.3

any program. There is even a theorem, called the Normal Form Theorem, which proves this.

A good flowchart has the property that all of its boxes and diamonds represent approximately the same level of detail. This is possible because a contiguous set of overly detailed symbols can always be grouped into a single more general box or diamond. Similarly, an overly general symbol can always be broken down into constituent symbols.

Flowcharts have enjoyed tremendous popularity, particularly as a tool for designing batch programs. For certain kinds of program they serve as an excellent intermediate stage between the concept and the code. A flowchart closely resembles what a computer actually does; so if you can draw a flowchart, you are guaranteed to have a design which can be translated into a program.

Despite this convenience, flowcharts have some shortcomings, and unfortunately these are most acutely felt in real-time design.

First, a flowchart tells nothing about the structure of the program it represents. By structure we mean the organization of the program into routines and subroutines, and the nature of the data which is passed between and among those routines and subroutines. The diagrammatic design language which addresses this need is the structure chart. We will discuss structure charts in the next section.

Secondly, a flowchart tells nothing about the flow of information through the system, which is a very different thing from the flow of control through the program. The design language which does this is the data flow diagram, which we will examine in Section 3.3.

The last shortcoming of the flowchart is a bit more subtle. Remember that the advantage of the flowchart is the way it mimics the action of a processor: orderly, step-by-step execution of one thing at a time. Real-time systems create the illusion of doing many things at once, and this property isn't very easily represented by a flowchart. The illusion is created by straightforward context switching and multiple process control techniques. The switching and the process control could of course be flowcharted, but the result would be a picture of the program's executive system (a.k.a., operating system) rather than a picture of the program as a whole.

Much of the magic of real-time systems comes from interrupts, and interrupts cannot be represented easily in a flowchart. The interrupt routine itself can be represented by a box, but where does the box go in relationship to the rest of the flowchart? That depends on when the interrupt occurs, and the interrupt could occur any time that interrupts are enabled. A first attempt might be to insert the following structure on every arrow of the flowchart (see Fig. 3.4).

This instantly makes the flowchart unreadable by quadrupling the number of arrows, and anyway it doesn't do the trick. As long as interrupts are enabled, an interrupt can be serviced between any two instructions,

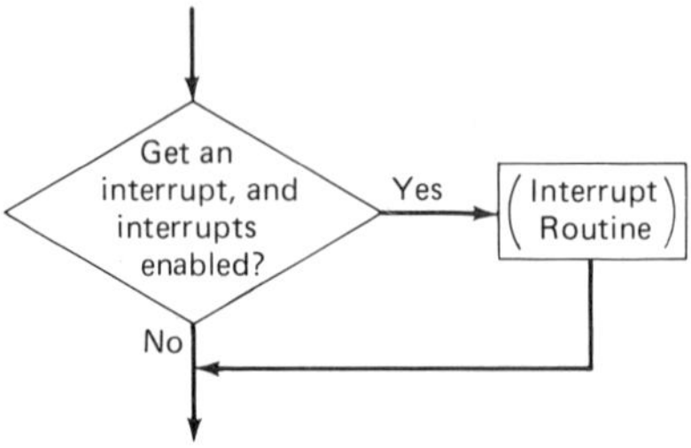

Figure 3.4

not just between any two flowchart symbols. We would have to put every instruction in its own box or diamond, and then we would have no advantage at all over an assembly language listing.

Because of these defects we will not use flowcharts in this book for high-level design. However, we will continue to use of them for representing lower-level processing, and you need not eliminate them from your vocabulary of high-level design techniques.

3.2. Structure Charts

In the last section we observed that one shortcoming of the flowchart is that it does not reveal anything about the structure of the program which it represents. In this section we will look at the structure chart, a diagrammatic design language which is specifically intended to reveal structure.

In investigating a program's structure we are interested in two things. First, we want to know about the hierarchy of subroutines: which routines and subroutines call which subroutines. Second, we want to know about the data which a calling routine passes to a subroutine, and also about the data which a subroutine returns to its caller.

Probably the structure chart first saw the light of day at I.B.M. in the early 1960's. Dissatisfied with flowcharts, I.B.M. wanted to develop a more sophisticated tool for design and documentation. The result was a set of guidelines called HIPO. HIPO is an acronym for Hierarchy, Input, Process, and Output, which are the four features it was intended to represent. Hierarchy was represented by the structure chart. Here we have somewhat modified the original concept to make it more appropriate for real-time software design. Please refer to Paige-Jones for an excellent orthodox description.

As with the flowchart, routines, subroutines, and processes are represented by boxes. Hereafter we refer to programs, routines, subroutines, and processes as "modules." The name of the module goes inside the box. Thus the program to write this chapter would be represented as in Fig. 3.5.

Figure 3.5

Suppose there were subroutines to write each section in this chapter. (Don't I wish there were!) The main program would call each of these subroutines. In a structure chart, a subordinate module is drawn below its caller, joined by an arrow pointing down (see Fig. 3.6).

Nothing is implied about the order in which the subordinate modules are called. In fact, a module may be called any number of times, even though it is only represented once. Some modules may never even be called at all. The structure chart only tells that somewhere in the calling module is a reference to the subordinate module.

Subroutines may be called by more than one caller. After all, that's the whole idea of subroutines. Thus it is permitted for several arrows from different boxes to enter a single subordinate box. For example, the first step in writing a section of a chapter is to write a title. A subroutine to write a section title would be called by all of the section-writing modules, as shown in Fig. 3.7.

Structure charts also show the parameters which are passed between calling and subordinate modules. Parameters are represented by small arrows with round tails, lying near and parallel to the arrow which joins the modules. The small arrow is labelled with the name of the parameter. If the parameter is passed to the subordinate module, the arrow points down (toward the subordinate); if the parameter is returned from the subordinate to the caller, the arrow points up (toward the caller). To take an example, suppose a routine named CALCULATE calls a subroutine named ROOT, with a parameter named NUMBER. ROOT calculates the square root of its input parameter, returning the result in a variable named RESULT. We would represent the relationship between CALCULATE and ROOT with the following structure chart (see Fig. 3.8).

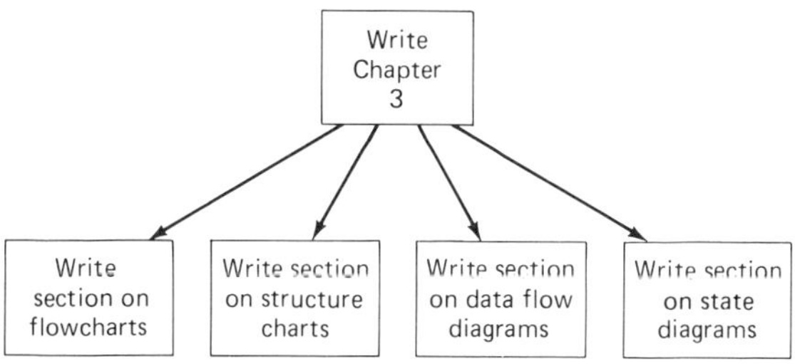

Figure 3.6

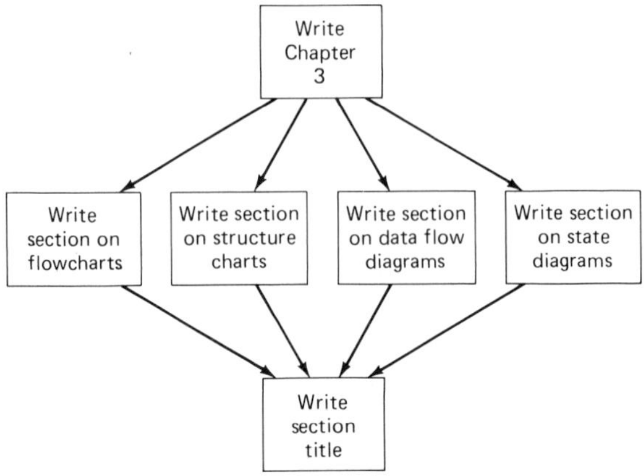

Figure 3.7

It is convenient to distinguish boolean parameters (true/false flags) from other kinds of parameters. A boolean variable is represented with the round tail of its arrow colored in. To extend our example, we would like ROOT to return an error flag called ERROR if the parameter passed to it is negative. Our structure chart would look like Fig. 3.9.

Let's take a look at a more detailed example. Imagine a typical video game in which the player maneuvers a ship on a display screen. The ship can rotate as well as translate. Also, the ship can fire missiles at an alien enemy. (In video games it is morally acceptable to kill intelligent life entities provided they are not human.) Because of hardware limitations, at most one missile can be displayed on the screen at any time. This gives rise to a rule: the ship may not fire until after its previous missile has hit a target or disappeared off the edge of the screen.

Such a game would be controlled by a microprocessor running a real-time program. In this program it would be useful to have a subroutine called GET STATUS which could return useful information about the ship and any missile which might be on the screen. The structure chart shown in Fig. 3.10 depicts GET STATUS.

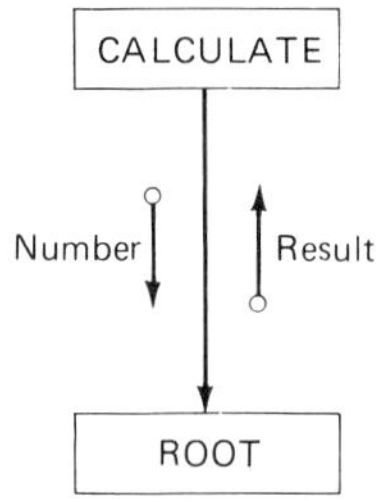

Figure 3.8

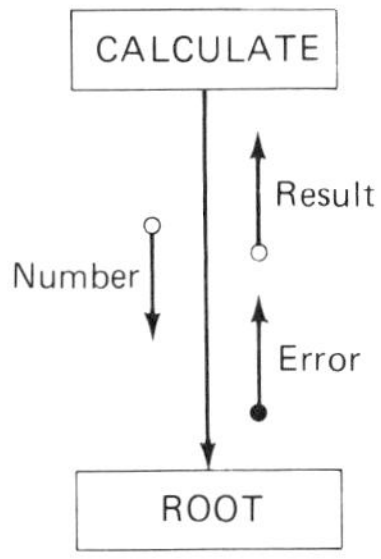

Figure 3.9

GET STATUS calls two subroutines, GET SHIP STATUS and GET MISSILE STATUS. Then GET STATUS bundles the ship and missile statuses into a record variable called simply STATUS, which it returns to its caller. (By "record variable" we simply mean a variable which, as in PASCAL, is composed of multiple pieces of information.)

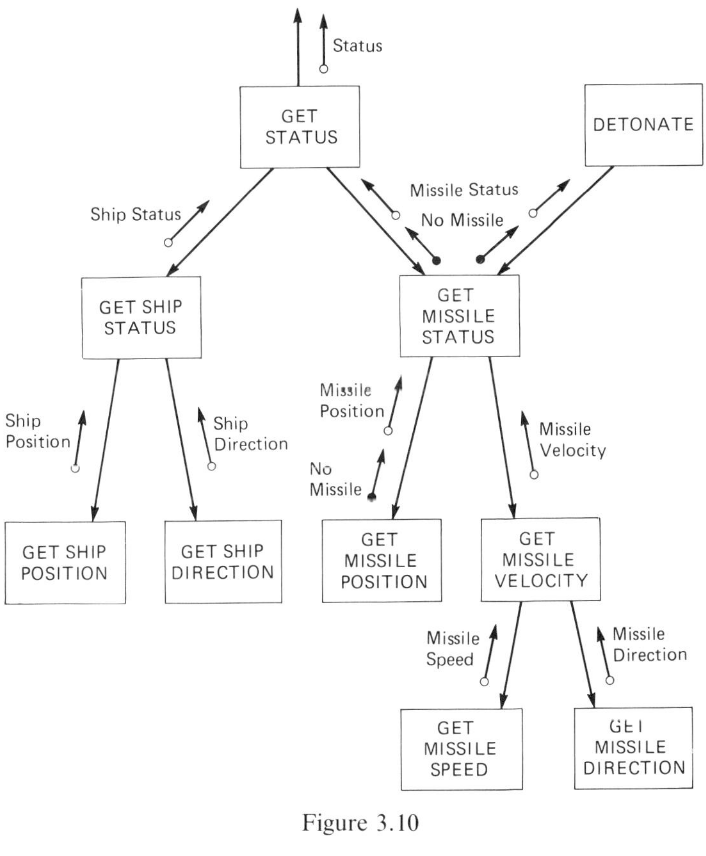

Figure 3.10

GET SHIP STATUS returns a parameter called SHIP STATUS, which is a record variable consisting of the ship's position and the direction it is facing. Subordinate subroutines are called to return these pieces of information.

GET MISSILE STATUS tells its caller whether or not there is a missile on the screen via the NO MISSILE flag. Moreover, if there is a missile, MISSILE STATUS is valid. MISSILE STATUS consists of the missile's position and its velocity; in turn, the velocity has a speed component and a direction component.

At this point it should be obvious that a structure chart is superior to pure prose for conveying this kind of information.

Note that GET MISSILE STATUS is also called by something called DETONATE, which draws a red explosion on the screen whenever a missile hits a target. It stands to reason that DETONATE would need to know about the status of the missile which it is detonating.

There might be no missile on the screen at the time when GET MISSILE STATUS is called. In this case, GET MISSILE POSITION will return with NO MISSILE set to TRUE. We see from the structure chart that GET MISSILE VELOCITY has no provision for the no-missile situation, so GET MISSILE STATUS should not call GET MISSILE VELOCITY if there is no missile.

This last observation illuminates one of the great virtues of structure charts: they tell us how to call subroutines safely. This is an infinite boon for somebody who has to modify the program without taking the time to learn its every facet. The structure chart tells us right away that GET MISSILE POSITION informs its caller if there is no missile, but GET MISSILE VELOCITY does not.

Suppose now that we discover that GET STATUS really doesn't need to return the direction the missile is travelling. In order to speed up the program, we would therefore like to modify GET MISSILE VELOCITY so that it does not call GET MISSILE DIRECTION. A glance at the structure chart tells us that we must be cautious. GET MISSILE STATUS is also called by DETONATE, so we should inspect DETONATE to see if it ever makes use of the missile's direction. If not, we can proceed as planned; but if so, we should abandon our plan and leave GET MISSILE VELOCITY alone. (Alternately, we could have DETONATE call GET MISSILE DIRECTION directly).

What if we want to rewrite GET SHIP POSITION so that SHIP POSITION is returned in a different format? The structure chart tells us that we can proceed without fear, because no other routine calls GET SHIP POSITION.

Our example has illustrated the principal benefit of designing and documenting with structure charts, namely that they allow the engineer to tell at a glance whether or not a modification may be made safely. Suppose we had wanted to modify GET MISSILE VELOCITY as before, but without the benefit of a structure chart. Either we would have to scan every line of the program source listing for a call to GET MISSILE VELOCITY, or we would make the change and hope that any problems introduced would manifest themselves clearly and benignly. Changes always have to be made to programs, both in their development phase and after their release. Any design tool which permits these changes to be made quickly and competently is worth its weight in gold.

We should point out one shortcoming of structure charts. They assume a clear division among the levels of the hierarchy of modules, and also a clear distinction between calling modules and subordinate modules. Things are not always so clear, particularly low-level time-critical streamlined high-performance code. As an example in 8Ø86 assembly language, suppose at the end of a subroutine a call is made to another subroutine named CALC. Suppose CALC looks like this:

```
CALC:  MOV AH,AL
       ADD AH,AH
       AND AL,AH
       XOR AL,55H
       RET
```

Now suppose that the last few lines of a subroutine which calls CALC look like this:

```
       ...
       MOV  SI,AX
       MOV  AL,7
       CALL CALC
       RET
```

What if we discover that system performance could be improved immensely if we sped up this routine. We could rewrite it as follows:

```
       ...
       MOV SI,AX
       MOV AL,7
CALC:  MOV AH,AL
       ADD AH,AH
       AND AL,AH
       XOR AL,55H
       RET
```

Functionally there is no difference, but a call (to CALC) and a return (from CALC) have been eliminated, thus providing the desired speed-up. This would be very difficult to represent in a structure chart, since it is ambiguous whether CALC is a module in its own right or just part of the routine in which it is imbedded.

Rather than conclude this section on a negative note, let us summarize by saying that structure charts are an excellent tool for design, development, and documentation.

3.3 Data Flow Diagrams

Of all diagrammatic design languages, the data flow diagram most closely captures the flavor of real-time systems.

The data flow diagram was originally a tool for systems analysis, used to portray the flow of information and material through a company or manufacturing operation. With some modification to the original notation, we can very successfully apply data flow diagrams to software. Once again, please refer to Paige-Jones for a fine orthodox description.

To introduce the symbols and ideas, let's start with an example which has nothing to do with computers. Please refer to the diagram in Fig. 3.11.

Imagine a candy-making machine. The product is simple: a chocolate bar with peanuts in it. The machine has three parts. First, a sheller takes unshelled peanuts and removes and throws away the shells; the shelled peanuts are temporarily stored in a bin. The second part of the machine takes melted chocolate, mixes in peanuts from the Shelled Peanut Bin, and pours the mixture into moulds. When the candy bars have hardened, they pass along a conveyor belt to a wrapping unit, which wraps them in preprinted wrappers from a bin. Lastly, the wrapped bars pass along a second conveyor belt and are deposited in a crate.

What observations can we make about this system and its accompanying data flow diagram?

First, we notice that everything is happening at once. While peanuts are being shelled in one part of the machine, candy bars are hardening in another part. Unshelled peanuts, liquid chocolate, and wrappers are constantly being imported. Finished goods and discarded peanut shells are constantly being exported. In Chapter 1 we listed some characteristics of real-time systems, and one of those characteristics was precisely this ap-

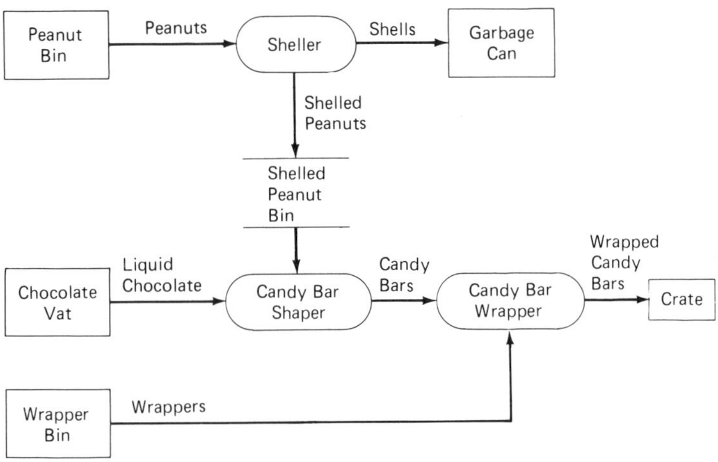

Figure 3.11

pearance of performing many functions simultaneously. Thus it is significant that the structure chart depicts this feature.

We can also observe that part of the system is for processing (the Sheller, the Candy Bar Shaper, and the Wrapping Unit), part is for transportation (the conveyor belts, for instance), and part is for storage (the Shelled Peanut Bin, the Chocolate Vat, and so on). The symbols of the structure chart differentiate among these functions. Bubbles represent processing, and as always the name of the process is written inside the bubble. Arrows represent transportation, and the name of the commodity being transported is written along the arrow. Boxes and parallel horizontal lines represent storage elements.

Why are there two symbols for storage elements? Notice in the data flow diagram that all of the boxes are on the periphery of the diagram, and conversely that the periphery consists of nothing but boxes. These boxes represent the system's inputs and outputs. The diagram does not show where the inputs come from or where the outputs go to, because those connections are not part of the system. Clearly unshelled peanuts must somehow get into the Peanut Bin, chocolate must be melted and put in the Chocolate Vat, and wrappers must be printed and stored in the Wrapper Bin; but the system doesn't care how this is accomplished. Similarly, when the Crate is full of finished candy bars it must be emptied or replaced, but again this is irrelevant to the analsis of the system at hand.

The Shelled Peanut Bin is represented by the second kind of storage symbol: parallel horizontal lines. This is because it is connected only to other parts of the system. It is filled by the Sheller and drained by the Candy Bar Shaper. It is a temporary internal storage unit, not an input/output device like the Chocolate Vat or the Garbage Can.

Now we can adapt the vocabulary of data flow diagrams to refer to real-time computer programs rather than to physical systems.

The boxes on the periphery of the system will represent peripheral I/O devices. This is logical, since peripheral devices constitute a system's connection with its environment.

Internal temporary storage corresponds to the mechanisms used by a system to store internal data. In large business systems, temporary disk or tape files usually play this role. Real-time systems generally have no disks or tapes; temporary data is stored in data structures in RAM. Thus for our purposes the parallel-lines symbol will be used to represent data structures.

Bubbles will still be used to represent processing, but instead of physical processing they will stand for data processing—in other words, code.

What about the commodities that flow along the arrows? In computer systems there is only one kind of commodity: data. The labels on the arrows will tell about the kind of data passing along the arrows between routines, data structures, and I/O devices. The arrows themselves have no direct counterpart. One might say that they represent the passing of parameters.

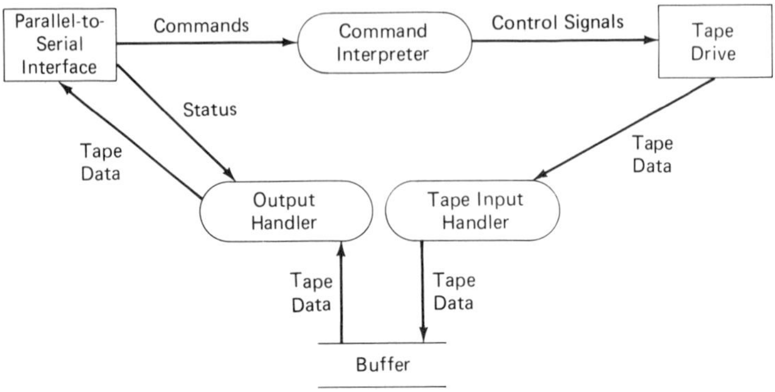

Figure 3.12

Now let's use a data flow diagram to depict a real-time program. Consider a simple system which interfaces to a magnetic tape drive. Commands from the outside world arrive at the system via a Parallel-to-Serial Interface. These commands tell the system to rewind, advance, or read the tape and report the data back along the Parallel-to-Serial Interface. We add the feature that data can be read from the tape faster than it can be written to the Parallel-to-Serial Interface.

The data flow diagram below portrays our system. There are three processing bubbles. The Command Interpreter reads bytes from the Parallel-to-Serial Interface and decodes them into control signals which are issued to the Tape Drive. The Tape Input Handler reads bytes from the tape. These bytes cannot be output directly to the Parallel-to-Serial Interface, so they are temporarily stored in the Buffer. The third process, the Output Handler, monitors the Parallel-to-Serial Interface, and outputs data from the buffer only when the interface is ready (see Fig. 3.12).

Notice how the data flow diagram has accurately communuicated the feeling of multiple processes happening simultaneously in real-time.

In the next chapter we will develop a technique for translating a data flow diagram into a program. Meanwhile, we can make the following important observation: since data flow diagrams nicely portray the activities of real-time systems, a good way to begin designing a real-time program is to draw a data flow diagram.

We will now briefly leave off our development of data flow diagrams because there is one last diagrammatic design language to discuss.

3.4 State Diagrams

We conclude our study of design languages with a look at state diagrams.

State diagrams have a fairly limited applicability, which is why we left them for last, but there are times when nothing but a state diagram will be useful.

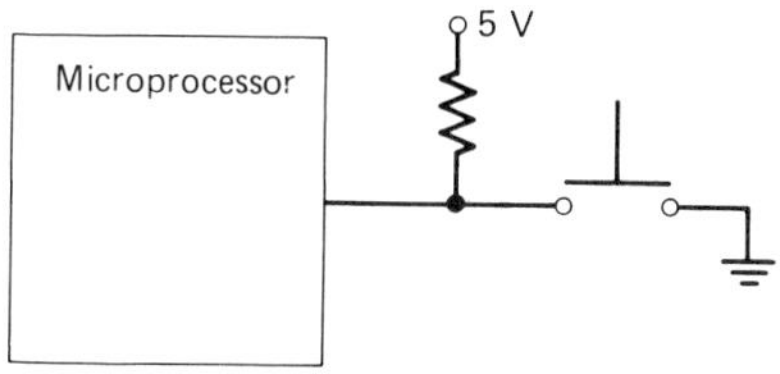

Figure 3.13

As a practical example, consider the problem of the Bouncing Switch. A routine in a microprocessor program is required to scan a single input line, which is connected to a pushbutton switch. When the switch is open, the input is in the TTl "1" state (approximately 5V). When the switch is closed, the input is in the TTL "Ø" state (approximately ØV). Whenever the switch switches, our routine has to update a variable named S to reflect the switch's new state. $S = 1$ will mean the switch is closed, and $S = Ø$ will mean it is open. We will assume that the routine will be called frequently, so that it can speedily update the value of S.

As a first attempt, suppose we try the routine shown in Figs. 3.13 and 3.14.

When the program is run, it is observed that the routine reports an inordinate number of switch transitions. In fact, every time the switch is closed, the processor reports that it closed, then opened, then closed, and so on, several dozen times.

The problem is that switches "bounce" when they close. Mechanical contact is briefly made when the button's plunger "hits bottom," but the

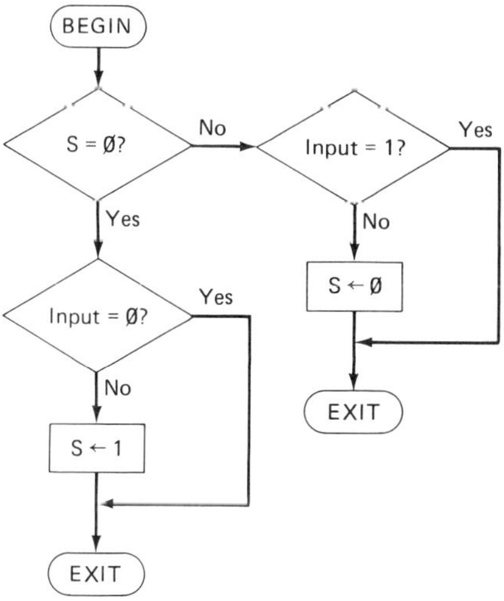

Figure 3.14

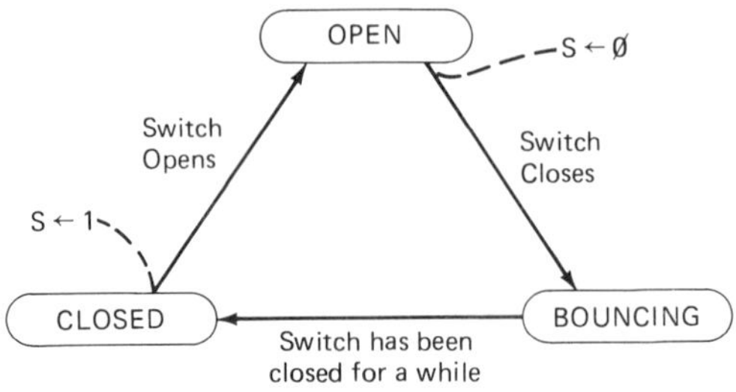

Figure 3.15

plunger immediately bounces up off of its contact, then returns, then bounces up again, and so on. Finally the switch comes to rest after a few milliseconds. The processor runs quickly enough, and the routine is called frequently enough, that many of the bounces are detected and mistakenly construed to be genuine transitions of the switch.

The problem with the flowchart is that it assumes that the switch has only two states, open and closed. In fact, there are three states: open, bouncing, and closed. We can depict the situation with the state diagram shown Fig. 3.15.

Each state is represented with a bubble, and arrows represent transitions between states. At any moment the system is in exactly one state. The circumstances which cause a system's state to change are written along the arrow which joins the old state to the new one. If the system takes some kind of action in the course of making a transition, the action taken is written near the appropriate arrow, joined to the arrow's tail by a broken line.

SWITCH HAS BEEN CLOSED FOR A WHILE is a very vague expression. How can we make it more concrete? Let's say that the bouncing lasts for a maximum of 5 milliseconds, so we will be safe if we wait 7 milliseconds after the switch closes. We could simply delay in a loop for 7 milliseconds, but this would deprive the rest of the program of a lot of valuable processing time. Instead, let's equip the system with a Counter/Timer which can be configured to interrupt after 7 milliseconds.

We need to introduce a new variable which will reflect the state of the pushbutton. We will call the variable STATE, and its possible values will be as follows:

Value of STATE	Meaning
Ø	Switch is open
1	Switch is bouncing
2	Switch is closed

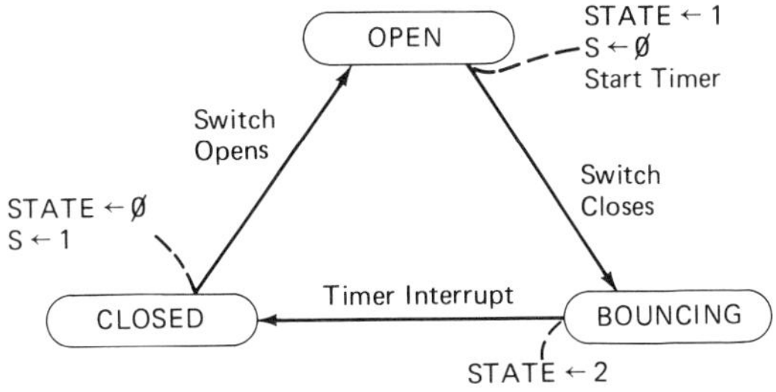

Figure 3.16

Naturally, when a transition is made, the value of STATE should change. The correct interpretation of the input signal from the pushbutton depends on the value of STATE. We need to update our state diagram a little bit (see Fig. 3.16).

Based on this state diagram, we can draw a correct flowchart (see Fig. 3.17).

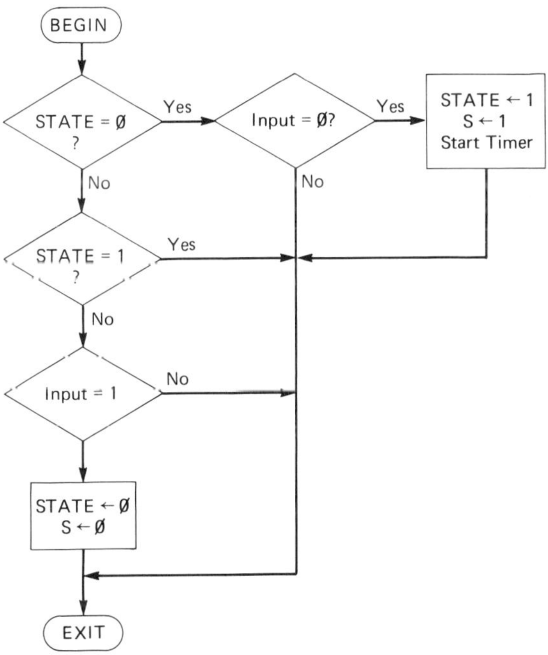

Figure 3.17

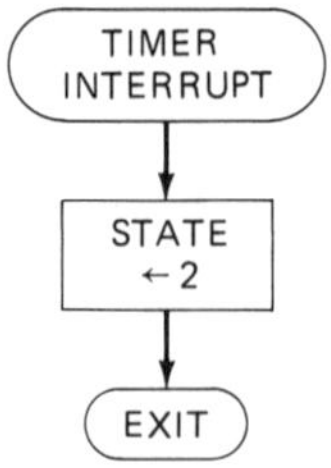

Figure 3.18

We also need a little interrupt routine (see Fig. 3.18).

State diagrams are useful when two conditions apply. First, when the proper response of a system to a stimulus depends not only on the stimulus itself, but also on some state of the system. Second, when part of the response may be to alter the state on which the response depends. In our example, the proper response to the stimulus of the switch opening depended on the value of the variable STATE. If STATE was 2, the proper response was to update S and change to state Ø; but if STATE was 1—that is, if the switch was bouncing—the proper response was to do nothing.

State diagrams are usually used to design very specific low-level portions of real-time programs. They are rarely appropriate for general top-level design. In the occasional instance when they are useful, however, they are enormously useful.

CHAPTER 4
Implementing the Data Flow Diagram

In this chapter we come to the prime point of our topic. We have already sung the praises of data flow diagrams. We are particularly interested in them because they portray so clearly the workings of real-time systems. However, they would be useless indeed if we had no technique for translating them into programs. In this chapter we develop such a technique.

Data flow diagrams show the dynamics of information as it is input, transported, stored, processed, and output. We will begin our discussion with a look at how to store data.

4.1 Circular Queues

We need a suitable data structure for storing real-time information. You can see from the title of this section that we will use circular queues. In fact, circular queues are practically the only appropriate data structures for our purposes. Look at why this is so.

Our first requirement of any data structure to correspond to the horizontal parallel lines of the Data Flow Diagram is that it preserve order. In other words, the order in which data is retrieved from the structure has to be the order in which it was originally stored. A teletype, for example, must transmit characters in the exact order they were typed. If the order is garbled, the teletype has not performed its function correctly. Clearly, if someone types in "DYSLEXIA" the teletype should transmit "DYSLEXIA" and not "DAILYSEX." In both cases the letters are the same, but you can see how important it is to get the order right.

The term for the order-preserving property which we desire is FIFO, which stands for "First In, First Out." The first byte stored in the data structure is the first byte retrieved. (And of course the second byte stored is the second byte retrieved, and so on.) "Queue" is a generic name for any data structure with the FIFO property. Since we require the FIFO property, we know that we are looking for some kind of queue.

Queues are accessed by subroutines for storing and retrieving data. The storing subroutine is usually called the PUT routine, and it is called with a parameter which is the byte to be put on the queue. (For the purposes of this discussion, we assume that we will always be storing and retrieving bytes. The principles we will develop can easily be extended to words or other units of information.) The retrieving subroutine is usually called the GET routine, and it returns a parameter which is the next byte retrieved from the queue.

If the queue is full—that is, if every location allocated for it in RAM is already storing information—then the PUT routine should return a flag to tell its caller that it cannot perform the put. Similarly, if there is no data in the queue, then the GET routine should return a flag to inform its caller that there is nothing to GET. We can express this with structure charts (see Fig. 4.1).

Queues are implemented as contiguous blocks of memory. Bytes are stored starting in the lowest address in the block; each byte is stored in the location immediately following the location where the previous byte was stored. To keep track of the end of the queue, there is a pointer, usually called the TAIL POINTER. The TAIL POINTER always points to (i.e., contains the address of) the next free location in the queue. The PUT subroutine stores its parameter in the location pointed to by TAIL POINTER, and then increments TAIL POINTER. If the PUT subroutine detects that TAIL POINTER points beyond the block allocated to the queue, it should immediately return with an error flag set. Please refer to the flowchart in Fig. 4.2 below. (The upward pointing arrow in the first line of the process box is an abbreviation for "the location pointed to by." Thus the first line in the box means "store the parameter in the location pointed to by TAIL POINTER.")

What about retrieving data? The GET subroutine needs a pointer, called the HEAD POINTER, which always points to the next byte to be read from the queue. The GET subroutine should retrieve the byte pointed to

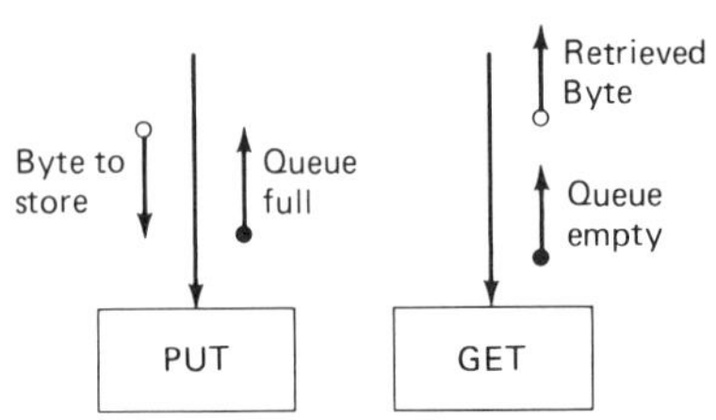

Figure 4.1

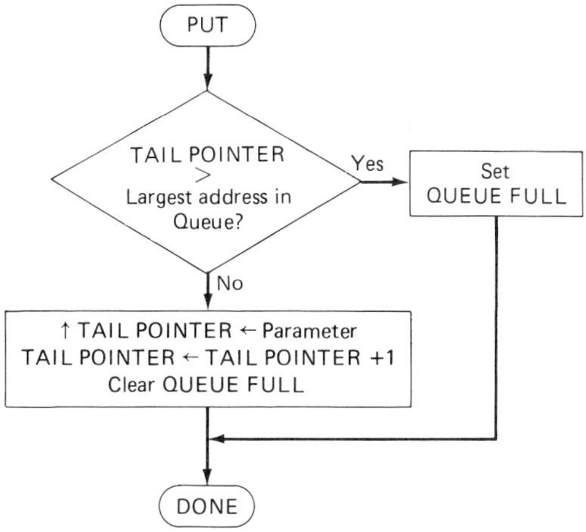

Figure 4.2

by the HEAD POINTER, and then increment the HEAD POINTER. If the HEAD POINTER and the TAIL POINTER are equal, then the queue is empty; in this case, the GET subroutine should set its error flag and return without altering the HEAD POINTER. The flowchart in Fig. 4.3 describes GET.

As an exercise in programming, let's code these two subroutines in 8Ø86 Assembly Language. We will use two variables called HEADPTR and TAILPTR for the two pointers. We will assume that there is a constant called PAST_QUEUE, which is the address of the first byte beyond the block allocated to the queue. We will use the AL register for passing

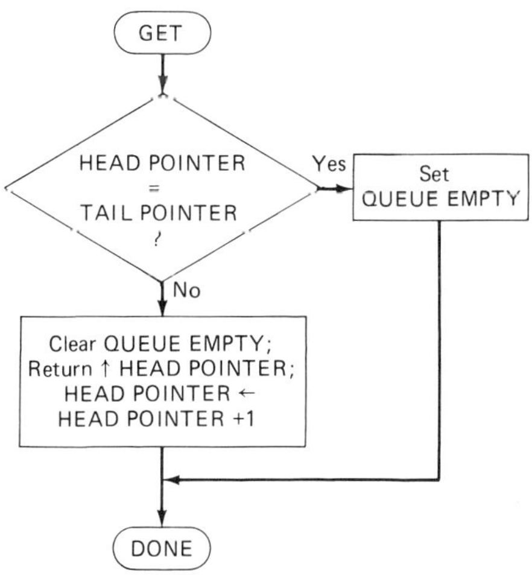

Figure 4.3

parameters. We could use other registers or variables in memory for the error flags, but it is more convenient to use the Carry Flag. If either subroutine returns with Carry set, an error has occurred. This makes life (life???) much easier for the calling routine: instead of testing a value in memory or in a register, the caller can perform a JC (Jump if Carry set) conditional jump instruction to jump to the appropriate error-handling code.

The two subroutines appear on the next page. PUT stores the byte in AL on the queue, GET returns with the next byte from the queue in AL. Both return with Carry set to indicate an error.

This is almost what we want, but not quite. The kind of queue we have described can only store a limited amount of data. When the last location allocated to the queue has been used, the queue is worthless. Even after the last byte has been retrieved and the queue is completely empty, PUT will always return with Carry set to indicate an error.

A queue with this defect is perfectly acceptable for applications in which the number of bytes to be stored is limited. Real-time systems, however, tend to run indefinitely, with no limit on the amount of data which might need to be queued, so the kind of queue we have described needs to be modified slightly.

Our queue resembles a line segment: it begins at one end and eventually terminates at the other end. We would like to store an unbounded amount of data. We can do this by closing the segment on itself, so to speak, to form a circle. The resulting data structure is called a Circular Queue.

With a Circular Queue, after data is stored in the last address, the next byte is stored back in the first address. Hopefully, the byte which was originally stored there will have been retrieved beforehand; otherwise an error will occur.

A circular queue can be used forever, provided that on the average it is emptied at least as fast as it is filled.

When designing Circular Queues, it is convenient to have a variable to tell how many bytes are in the queue. Let us call this variable "SIZE." We also need to know the number of bytes of storage allocated to the queue. For the example below, let's assume that 750 bytes are allocated. Let's assume that the queue begins at a location labelled "QUEUE_START." At a typical moment the queue might look like Fig. 4.4.

A flowchart for the PUT routine might look like Fig. 4.5.

And the GET routine might look like Fig. 4.6.

Note how the HEAD POINTER and TAIL POINTER "wrap around" to the beginning of the queue after they reach the end.

Before we begin coding, let's look at what happens to the pointers when they get updated. Essentially, each pointer's distance into the queue is incremented, modulo 750. It happens that computationally it is very easy to increment in certain modulos—namely, those modulos which are powers of 2. In modulo 32, for example, we are only interested in numbers which are five bits long. Thus to increment the contents of a register mod-

```
PUT:      MOV SI,TAILPTR          ; SI → TAIL OF QUEUE
          CMP SI,PAST_QUEUE       ; QUEUE FULL?
          JE  PUT_ERR             ; YES, ERROR
          MOV [SI],AL             ; NO, STORE AL IN QUEUE
          INC SI                  ; UPDATE TAIL POINTER
          MOV TAILPTR,SI
          CLC                     ; RETURN WITH CARRY CLEAR
          RET
PUT ERR:  STC                     ; ERROR! RETURN WITH CARRY SET
          RET

GET:      MOV SI,HEADPTR          ; SI → HEAD OF QUEUE
          CMP SI,TAILPTR          ; QUEUE EMPTY?
          JE  GET_ERR             ; YES, ERROR
          MOV AL,[SI]             ; NO, AL ← NEXT BYTE
          INC SI
          MOV HEADPTR,SI          ; UPDATE HEAD POINTER
          CLC                     ; RETURN WITH CARRY CLEAR
          RET
GET_ERR:  STC                     ; QUEUE FULL, RETURN WITH
          RET                     ; CARRY SET
```

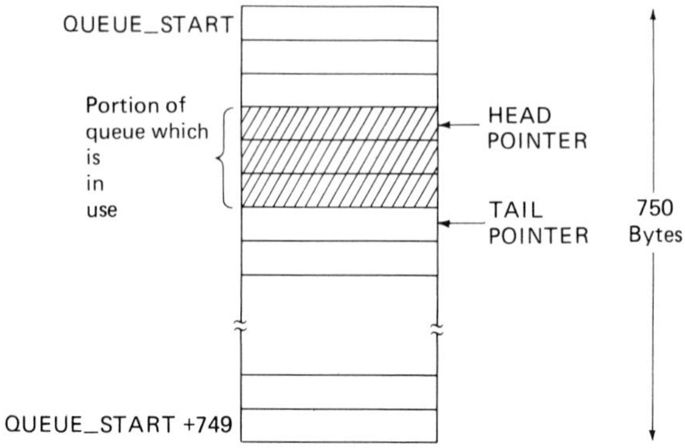

Figure 4.4

ulo 32, increment the register and then discard all but the lowest five bits by AND'ing with 11111 (binary). In general, to increment a register modulo 2^n, first increment and then AND with 2^n-1.

So the most convenient size for a Circular Queue is a power of 2. In the code below, we will use queues which are 512 bytes long. The HEAD POINTER and the TAIL POINTER will represent offsets into the queue, rather than absolute addresses. Thus these pointers will range from Ø through 511. Once again, parameters are passed in AL and the Carry flag indicates trouble. The program appears on the next page.

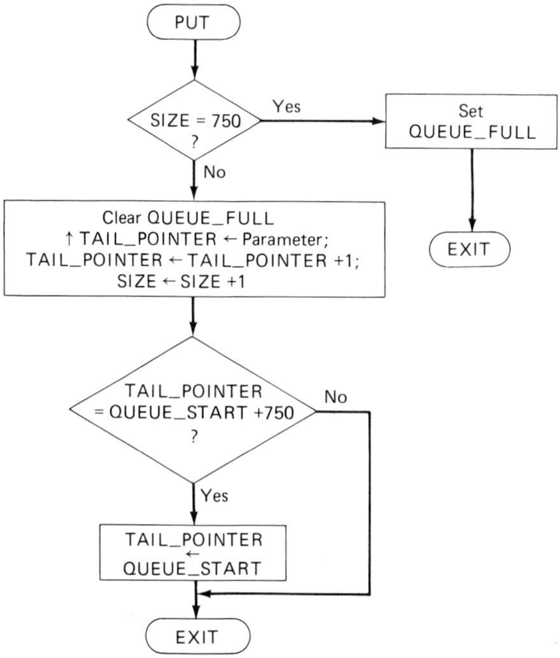

Figure 4.5

```
PUT:      MOV BX,QUEUE_ SIZE
          CMP BX,512                  ; QUEUE FULL?
          JE  PUT_ERR                 ; YES
          INC BX                      ; NO, INCREMENT SIZE
          MOV QUEUE_SIZE,BX
          MOV SI,TAILPTR
          MOV .QUEUE_START[SI],AL     ; STORE BYTE IN QUEUE
          INC SI
          AND SI,511                  ; UPDATE TAIL POINTER,
          MOV TAILPTR,SI              ; MOD 512
          CLC                         ; RETURN WITH CARRY CLEAR
          RET
PUT_ERR:  STC                         ; QUEUE FULL,
          RET                         ; RETURN WITH CARRY SET

GET:      MOV BX,QUEUE_SIZE
          CMP BX,Ø                    ; QUEUE EMPTY?
          JE  GET_ERR                 ; YES, ERROR
          DEC BX                      ; NO, UPDATE SIZE
          MOV QUEUE_SIZE,BX
          MOV SI,HEADPTR
          MOV AL,.QUEUE_START[SI]     ; READ NEXT BYTE INTO AL
          INC SI
          AND SI,511                  ; UPDATE HEAD POINTER,
          MOV HEADPTR,SI              ; MOD 512
          CLC                         ; RETURN WITH CARRY CLEAR
          RET
GET_ERR:  STC                         ; QUEUE EMPTY,
          RET                         ; RETURN WITH CARRY SET
```

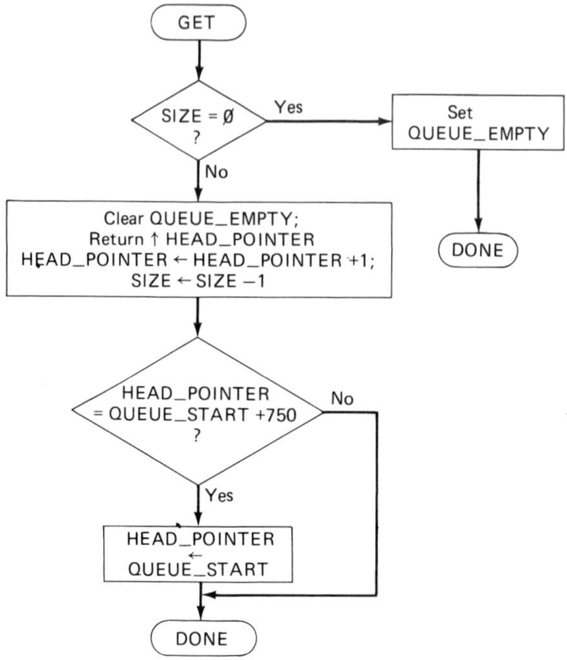

Figure 4.6

There are many ways to optimize Circular Queues. We will not discuss them here; we will only observe that in any particular program there is usually one best implementation for each Circular Queue, and the strategy for arriving at this implementation depends heavily on context.

4.2 Virtual Machines and Multiple Processes

It is now time to attack a subject which lies at the very heart of real-time programming. We were especially attracted to Data Flow Diagrams because they conveyed the feeling of everything happening at once. In this section we look at just what it is that is happening all at once. To do this we have to introduce the concept of Virtual Machines.

It is easiest to explain the ides of Virtual Machines with some examples from the popular media.

In the movie *Copacabana*, Groucho Marx plays an unscrupulous agent who represents a beautiful young singer, played by Carmen Miranda. In an effort to make as much money as possible, Groucho books his client to be in two different shows, in two different but adjacent night clubs, at the same time, under two different stage names. Carmen Miranda sings on one stage, frantically changes her costume, and runs to the second club. On her break, she changes clothes again and runs back to the first club, and so on. At one point, exasperated, she asks Groucho, "How can

I be in two places at once?'' Groucho replies, ''Boston and Philadelphia are in two places at once!''

This motif can be found elsewhere in our modern folk literature. As a boy I remember reading a Superman comic book in which Superman entertains children at an orphanage by playing ping-pong against himself. He serves, then runs at super-speed to the other side of the table to return. Then he runs back to the first side to return the return, and so on. He hits the ball at normal human speed, but he runs so quickly that he is invisible. The orphans think they see two identical Supermen, one on each side of the table.

A last example, although the reader has probably caught on to the pattern by now, comes from modern physics. It has been proposed, although I cannot say how seriously, that there is only one electron in the universe! This electron is shared equally among all things, moving here and there not only in space but also in time to give us the impression that the universe is populated with a staggering number of ''virtual electrons.''

By extension of the term ''virtual,'' we could say that in *Copacabana* the two night clubs each featured a virtual Carmen Miranda. In the Superman episode, the children saw two virtual Supermen playing ping-pong.

A user communicating via a terminal with a multi-user mainframe computer is oblivious to other users on the system, in the same way that the audience at one club in *Copacabana* is oblivious to the existence of the virtual singer at the other club. There is only a vague sense, in both examples, that things might be running a bit faster.

The various individual users of a computer are in fact each communicating with a dedicated VIRTUAL MACHINE. On the average, they each get their fair share of the computer. The definition of ''fair'' is implicit in the operating system, which activates and deactivates the virtual machines. The computer spends a certain amount of time being various virtual machines for its various users, and a certain amount of time switching personalities.

How does one computer support a number of virtual machines? First, each virtual machine is given its own portion of memory. Into this memory is loaded the program which the virtual machine is to run, and in it are stored all of the program's variables and data structures. For each machine which is waiting to be animated there is a data structure which represents the machine's state—that is, the contents of its registers, including the address of the next instruction to be executed when the machine is activated. This scheme works perfectly, provided all the virtual machines stay out of one another's memory.

A piece of code which executes on a virtual machine is called a PROCESS. A system which supports multiple virtual machines is, at the highest level, a manager of processes. Much work, both practical and theoretical, has been done in the study of process management. The best way to get acquainted with the subject is to take a course in Operating Systems.

In this book we will only look at process management in the context of real-time systems. Processes need to be treated somewhat differently in real-time systems than in general-purpose computer systems.

In a real-time system, what constitutes a process? If we can answer this question we will be well on our way toward establishing a methodology for implementing real-time processes.

Please refer to Fig. 4.7, which is a Data Flow Diagram for a fictitious real-time system. (This system is so fictitious that we have not bothered to give the boxes realistic names. We have also left out the data type names which usually appear along the arrows.)

All we know about this system is that A is an input device; W, X, Y, and Z perform data processing; 1 and 2 are circular queues; and B and C are output devices.

Since W, X, Y, and Z perform data processing, it is natural to look at them when we identify the processes in the system. We have no rigid guidelines for determining what should constitute a process, but we would like to preserve the intuitive notion that one process should be responsible for each (seemingly simultaneous) activity of the system.

Should each data-processing bubble be a process in its own right? This is a good start, but we would then have to have separate processes for Y and Z, even though they both seem to be engaged in the job of draining queue 2 and delivering output to device C.

We can get a better guideline if we look not at data processing, but at data. There are three sources of data in the diagram: input device A, and circular queues 1 and 2. Data from queue 1 is read by X and interpreted into output signals for output device B. Data from queue 2 is read by Y and processed; in reaction to this processing, Y passes data to (that is, calls) Z, which constructs signals for output device C.

This data-oriented description leads to the suggestion that there should be three processes: one to read data from A and take appropriate action; one to read data from 1 and take appropriate action; and one to read data from 2 and take appropriate action.

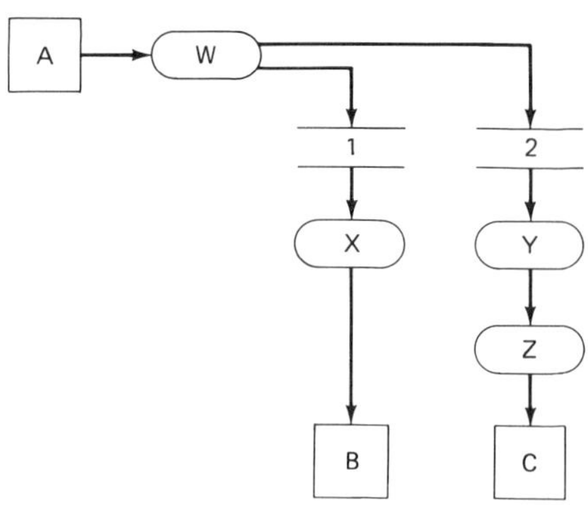

Figure 4.7

This leads us to propose the following generalization:

> In a real-time system, there should be a process for each source of data (input decives or circular queues). Each process should move data along until it reaches its destination in a Circular Queue or at an output decice.

This is only a rule of thumb, not a rigid definition, but it is an extremely useful rule of thumb.

4.3 Multiple Processes on the 8Ø86

It is a worthwhile exercise to implement a general scheme for managing processes on the 8Ø86. It should be kept in mind that any time consumed by the process manager is time which is taken away from the processes. Thus the scheme we will develop here will probably be too fancy to be incorporated in any real-time systems. The purpose of this section is to get the reader comfortable with the notion of processes, rather than to provide ready-made code.

Our goal is to develop a way for a process to "suspend" itself, yielding the processor to another waiting process. We would like to have a subroutine which any process can call, which would suspend the calling process and activate the next dormant process.

Having a subroutine to do this job may feel counter-intuitive. Process management is the highest-level function of the program, and yet the Process Manager is a subordinate subroutine. There is a good reason for this: although no process may run without permission from the Process Manager, once a process is activated it has full possession of the computer. The Process Manager cannot regain control on its own; the active process must graciously yield control by calling the Process Manager. This accounts for the Manager's dual nature as top-level dictator and subordinate subroutine.

(In multi-user computer systems, a mechanism is actually provided for the Process Manager to regain control. This is most often done with a Counter/Timer. When a process is activated, the Counter/Timer is set to interrupt after a certain interval; the length of the interval is generally a function of the priority of the process. If the process does not voluntarily suspend itself by the time the Counter/Timer interrupts, it is forcible suspended by the interrupt routine. Such schemes, while effective for general-purpose computers, are usually inapopropriate for real-time systems.)

When a process is suspended, its entire status must be saved, so that when it is reactivated it can continue as if nothing had happened. What exactly is the status of a process?

First, all of the registers must be preserved. This includes the IP (Instruction Pointer) and the CS (Code Segment), because we need to know what address in the process' code to jump to on reactivation.

Does anything else need to be saved? No. Processes store data in data segments and on stacks, but we assume that they do not unwittingly overwrite one another's data. So stored data, even though it is part of the status of a process, does not need to be preserved.

In the code that follows, we will assume that each process runs in its own code segment, stores data in its own data segment and extra segment, and even has its own stack in its own stack segment. After all, we are aiming for generality.

Before we continue, we need to introduce some 8086 instructions. We did not cover them in Chapter 2 because we did not want to flood the reader with descriptions of instructions without some example to relate them to.

Our instructions are PUSH, POP, PUSHF, POPF, CALLF, and RETF. They all have something to do with the stack. Recall that SS:SP (that is, 16*SS + SP) is always the address of the top of the stack. "Pushing" refers to decrementing SP and writing a 16-bit value to the new top of the stack. "Popping" refers to retrieving a 16-bit value from the top of the stack and then incrementing SP. Thus SP always points to the topmost item in the stack.

PUSH is the instruction which pushes a 16-bit value onto the stack. The syntax is:

```
PUSH  reg
```

where "reg" is the name of any 16-bit register. After the push, the register is not altered. Its contents are simply duplicated onto the stack.

POP is the instruction which retrieves data from the top of the stack. The syntax is:

```
POP  reg,
```

where "reg" is the name of the 16-bit register into which the 16-bit value on the top of the stack will be copied. Please note that "POP CS" is not allowed. "POP IP" is the same as "RET."

PUSHF is like PUSH, but there is no register specified. PUSHF pushes the values in all of the one-bit flags, such as Carry and Zero. POPF, which also has no parameter, pops values from the top of the stack into the flags.

CALLF is used for jumping to a subroutine in a different code segment. The values in both CS and IP are pushed (ordinary subroutine calls only push IP) and CS and IP are loaded with new values specified by the instruction. Different assemblers have different syntaxes for the argument to CALLF, most of them cumbersome. We will assume that if we give a subroutine's name as an argument to CALLF, then our imaginary assembler will provide the correct values for both CS and IP. Thus our syntax will be:

```
CALLF  label
```

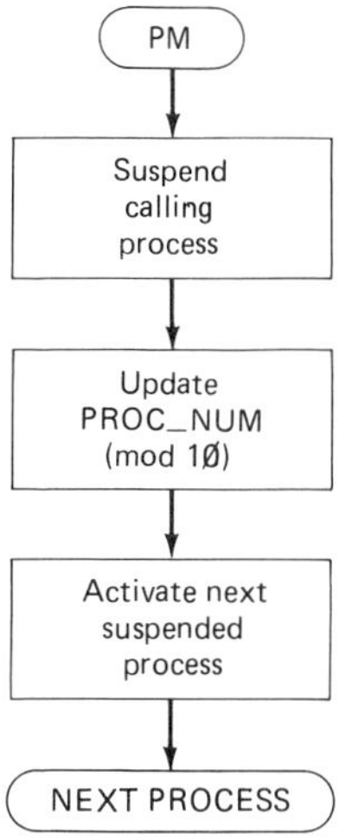

Figure 4.8

where "label" specifies the subroutine. The "F" in CALLF stands for "Far," because the range of the call is unlimited, whereas the range of an ordinary or "near" call is limited to the 64-Kbyte current code segment.

RETF is used for returning from a subroutine which was called using CALLF. RETF loads both IP and CS from the stack.

With our increased vocabulary of 8Ø86 instructions, we are ready to delve into code.

Since each process has its own stack, it is most convenient to store nearly all of a suspended process' registers on its own stack. We will store the values of SS and SP in the Process Manager's data segment, because the Process Manager needs to be able to find the stack of every suspended process.

We will use a round-robin priority scheme. This means that we will activate each process in turn; when the last process is suspended the first one is activated, and so on, round and round forever or until the machine is turned off.

The Process Manager will be a subroutine called PM (after all, it is the

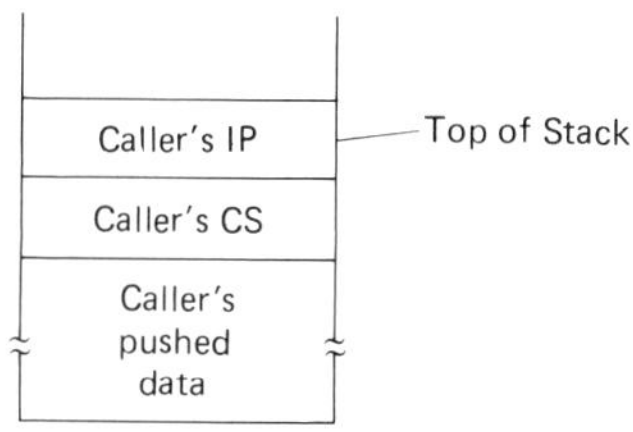

Figure 4.9

Prime Minister of the program). PM does not need a stack or a stack segment of its own, but it does need a small data segment. This segment might look like this:

```
PROC_NUM      RB 1
STACKS        RW 2Ø
```

We will assume arbitrarily that there are ten processes. The variable PROC_NUM contains a number from Ø to 9 which identifies the currently executing process. STACKS contains the SS and SP values for all of the suspended tasks, in order. First will be SS for process #Ø, then SP for process #Ø, then SS for process #1, and so on. We need two words for each of ten processes, so we have allocated twenty words in all at STACKS. These values will have to be initialized when the system is turned on.

We will also assume that PM's data segment begins at location ØØØØØØ.

A process which wishes to suspend itself should execute the instruction "CALLF PM". PM must do three things, as in the flowchart in Fig. 4.8.

The first part of PM suspends the calling process. Just as PM begins to execute, the caller's stack looks like Fig. 4.9.

PM begins by pushing nearly every register—in fact, everything except CS, IP, SS, and SP. CS and IP are already on the stack, because the caller did a "CALLF"; SS and SP don't need to be pushed because they will be stored in PM's data segment instead.

The first part of PM should look like this:

```
PM:  PUSHF
     PUSH  AX
     PUSH  BX
     PUSH  CX
     PUSH  DX
     PUSH  SI
     PUSH  DI
     PUSH  BP
     PUSH  DS
     PUSH  ES
```

After these instructions are executed, the caller's stack looks like Fig. 4.10.

Now we need to store the caller's SS and SP in PM's data segment. We may use any of the pushed registers to accomplish this. We will multiply PROC_NUM by 4 (by doubling it twice) and use the result as an index into STACKS.

```
MOV AX,Ø
MOV DS,AX                        ; DS → PM'S DATA SEG
MOV SI,OFFSET STACKS
MOV BL, PROC_NUM
```

```
        ADD  BL,BL
        ADD  BL,BL                 ; SI+BX → SAVE AREA FOR
        MOV  BH,Ø                  ; CURRENT PROCESS
        MOV  [SI+BX],SS            ; STORE CALLER'S SS
        MOV  2[SI+BX],SP           ; STORE CALLER'S SP
```

The calling process is now suspended. The next job is to update PROC_NUM, modulo 1Ø:

```
        MOV  AL,PROC_NUM
        INC  AL
        CMP  AL,1Ø                 ; Z FLAG SET IF AL=1Ø
        JNZ  PM_5                  ; IF AL<1Ø
        MOV  AL,Ø                  ; IF AL=1Ø THEN AL←Ø
PM_ 5:  MOV  PROC_NUM,AL
```

Lastly, we need to activate the next process :

```
        MOV  BL,PROC_NUM
        ADD  BL,BL
        ADD  BL,BL                 ; BH IS STILL Ø
        MOV  SI,OFFSET STACKS
        MOV  SS,[SI+BX]            ; SS ← NEXT PROCESS' SS
        MOV  SP,[SI+BX]            ; SP ← NEXT PROCESS' SP
```

Now we can pop register values from the new stack:

```
        POP  ES
        POP  DS
        POP  BP
        POP  DI
        POP  SI
        POP  DX
        POP  CX
        POP  BX
        POP  AX
        POPF
```

At this point the stack of the new process looks like Fig. 4.11.

A simple RETF will now clear up the stack and jump to the instruction immediately following the "CALLF PM."

Here is PM in consolidated form, slightly changed to take advantage of the fact that all of PM's three portions use PROC_NUM:

```
PM:     PUSHF                      ; PUSH FLAGS AND
        PUSH   AX                  ; ALL REGISTERS
        PUSH   BX                  ; ONTO CALLER'S STACK
        PUSH   CX
        PUSH   DX
```

```
        PUSH  SI
        PUSH  DI
        PUSH  BP
        PUSH  DS
        PUSH  ES
        MOV   AX,Ø
        MOV   DS,AX                 ; SELECT PM'S DATA SEG
        MOV   AL,PROC_NUM
        MOV   BL,AL
        ADD   BL,BL
        ADD   BL,BL
        MOV   BL,Ø                  ; SI+BX → SAVE AREA
        MOV   SI,OFFSET STACKS      ; FOR CALLING PROCESS
        MOV   [SI+BX],SS            ; STORE CALLER'S SS
        MOV   [SI+BX],SP            ; AND SP
        INC   AL                    ; UPDATE PROCESS # MOD 1Ø
        CMP   AL,1Ø
        JNZ   PM5
        MOV   AL,Ø
PM5:    MOV   PROC_NUM,AL
        MOV   BL,AL
        ADD   BL,BL
        ADD   BL,BL
        MOV   SS,[SI+BX]            ; LOAD NEW SS AND SP
        MOV   SP,2[SI+BX]
        POP   ES                    ; RESTORE NEXT PROCESS'
        POP   DS                    ; REGISTERS
        POP   BP
        POP   DI
        POP   SI
        POP   DX
        POP   CX
        POP   BX
        POP   AX
        POPF                        ; AND FLAGS
        RETF                        ; GO TO NEXT PROCESS
```

With this implementation of the Process Manager, each process is an infinite loop which calls PM at least once every iteration. The general flowchart for such a process could look like Fig. 4.12.

We use the term "dispatch" to mean "process a byte from its source through to its ultimate destination in a queue or at an output device." If dispatching a byte requires an inordinately long time, thus depriving the other processes in the system, the process can have calls to PM sprinkled at appropriate intervals.

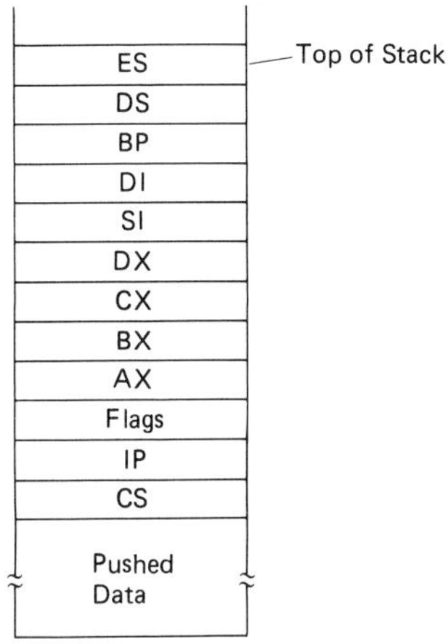

Figure 4.10

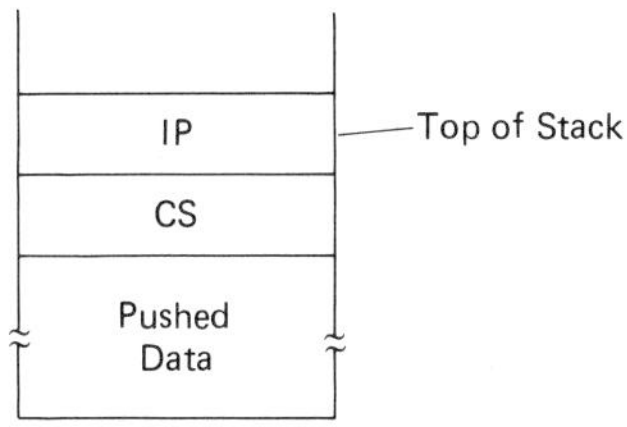

Figure 4.11

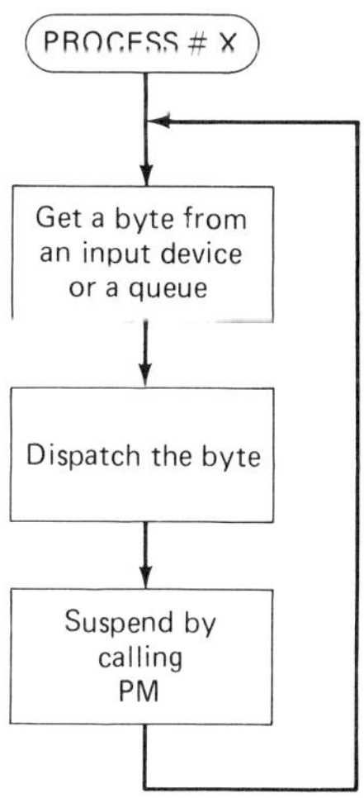

Figure 4.12

4.4 When Microseconds Count

In the worst case, it takes 376 clock cycles to call and execute PM. If the processor is running at 5 MHz, each cycle is 1/5 of a microsecond. It would take 75.2 microseconds to swap processes using PM as defined in the last section.

This may not be a problem, but in practice it very well could be. Those 75.2 microseconds are pure overhead. If the system in question is going to be marketed commercially, then it will be competing against similar products. Speed is a very important factor in the overall performance of the system, so even a small reduction in speed could mean the difference between a winning product and a losing one.

Moreover, if a system spends more time on overhead than is absolutely necessary then the hardware, and specifically the processor, is not being used to its maximum capability. The cost of the hardware is passed on to the customer. So if the program is inefficient, the buyer's money isn't going as far as it could.

Thus making code more efficient improves the legendary "price/performance ratio" on both sides of the division sign. Since the Process Manager is being called constantly, it is an obvious place to start making speed improvements.

We stated at the beginning of the last section that developing PM as we did was for the purpose of familiarizing the reader with the concept of processes. Having done so, we will now make the ultimate optimization of the process managing chore by *Eliminating the Process Manager Completely!*

Most of the Process Manager's time is taken up storing and restoring registers. Some of these registers may not contain meaningful data at the time PM is called, so pushing and popping them is a waste of time. Is there any way to know when a register does not have to be saved? Generally speaking, there is not. However, at the moment when a process has completely finished dispatching a byte, it is usually safe to say that NONE of its registers contains useful data. This is because, from the point of view that processes exist merely to dispatch data, the registers serve the purpose of helping to move and transform the original byte from the original input device or queue. Once the moving and transforming are finished, the registers have served their purpose and are no longer needed.

If this is indeed the case, then at the moment when the process has finished dispatching, it can be safely suspended without saving any registers at all.

But we observed in the last section that suspending a process means nothing more nor less than saving its registers, and activating a process means restoring its registers. We seem to have eliminated the need for the Process Manager altogether!

Processes can be implemented as subroutines which return when they have finished dispatching. Instead of a Process Manager, all we need is

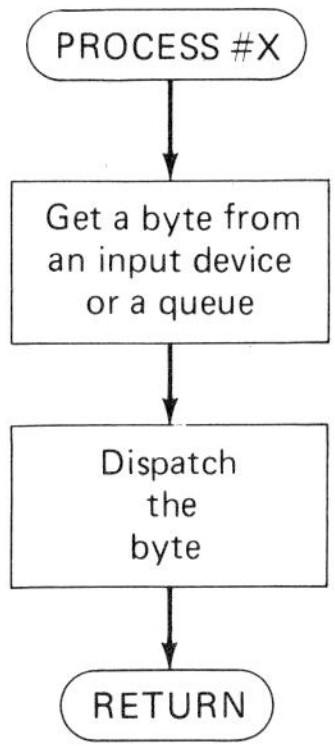

Figure 4.13

a master top-level routine which will call each process subroutine in turn.

We can now propose the model in Fig. 4.13 for processes in a streamlined real-time program.

Figure 4.14 gives us a model for the program's top-level routine.This is the streamlined high-performance real-time program in its purest form.

In practice, some processes may not present suspension-points where no registers need to be saved. Specifically, under some input circumstances a process may need so much time to process a byte that overall system performance is noticably degraded. In such a case, it may be necessary to suspend the offending process halfway through, even though some registers may need to be saved. This can be done as a special case, without

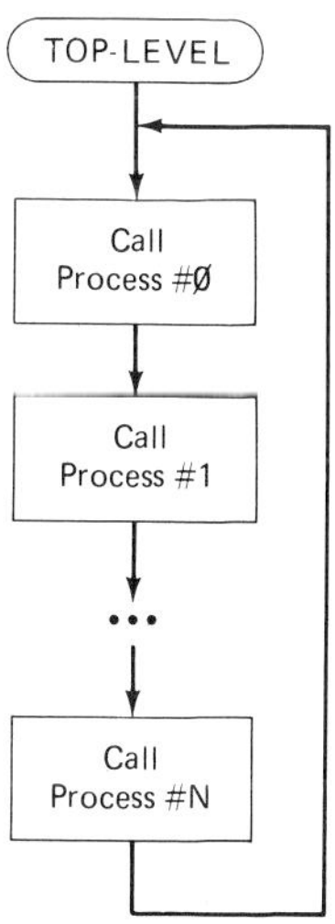

Figure 4.14

requiring that all of the other processes save their registers when they are suspended as well.

We have developed a general methodology for turning a Data Flow Diagram into a program. The method calls for implementing temporary data storage structures as Circular Queues, and creating processes for emptying the queues and the system's input devices. In the next chapter we will test this methodology by applying it at length to a practical example.

CHAPTER 5
A Detailed Example

It is now time to put all the theory of the previous four chapters into practice. We will work through one detailed example from start to finish, applying all the techniques we have learned up until now.

We will develop a complete 8086 assembly language program. For two reasons, this program will *not* be particularly optimized. First, in order to optimize assemble code for any processor, one must first be thoroughly familiar with every instruction supported by the processor, and it would be beyond the scope of this book to present every 8086 instruction. Moreover, it is important to have a concrete strategy for optimization, and we are going to defer discussing such strategies until the next chapter.

5.1 Definition of the Problem

We are going to design and write a real-time program to control a microprocessor-based intelligent keyboard. The keyboard is the human interface to a system whose nature is irrelevant to our discussion. The keyboard communicates serially with the system, as shown in Fig. 5.1.

The keyboard is fully decked out with fancy features. To begin with, there are 100 alphanumeric and control keys. Some of the keys, for example the CAPS LOCK key, are equipped with little L.E.D.'s which can be turned on and off under processor control. Thus the CAPS LOCK L.E.D. could be turned on when the system is in CAPS LOCK mode.

The keys are scanned by special "black box" scanning circuitry. Whenever a key moves up or down, the key-scanning hardware interrupts the processor. A unique 7-bit code identifying the key (known as the Key Address) is available at a certain input device, and the state of the key (whether it just went up or just went down) is available at another input device. After interrupting, the scanning hardware stops scanning until the

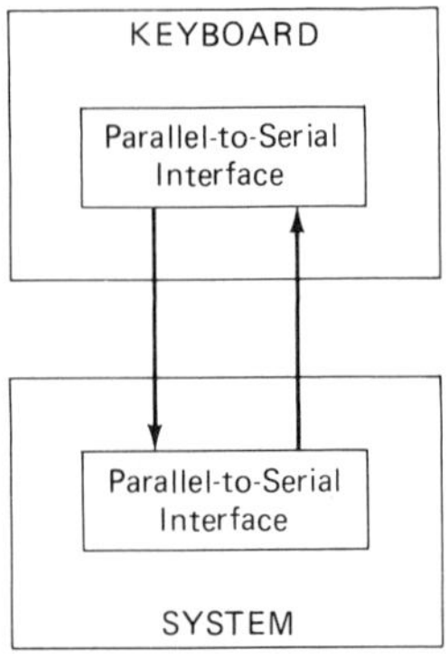

Figure 5.1

key state is read. It is important to read in the key state as soon as possible after the interrupt, so that subsequent key strokes will not be missed.

The keyboard hardware has a Parallel-to-Serial Interface for communicating with the system. In general, the system will send commands to the keyboard and the keyboard will send reports about key movement to the system. A Counter/Timer provides baud rate timing to the Parallel-to-Serial Interface.

A last portion of the hardware is a bell circuit. When a "1" is written to the output address corresponding to this circuit, an electronic beep is generated. The beep stops when a "Ø" is written to the output address.

Figure 5.2 shows a block diagram of the keyboard hardware.

When a key moves, a one-byte report called a "Keystroke" should be sent to the system. The 7 less significant bits of the byte should be the Key Address, and the most significant bit should be the key's new state ("1" for down, "Ø" for up).

We would like the keyboard to support the "auto repeat" feature: if any key is held down for more than 1/4 of a second, a new report that it went down should be sent to the host every 1/1Ø of a second. These reports repeat until either the key goes up again or another key goes down. Since this function involves timing some intervals, you can guess that we will be using the Counter/Timer.

The system gives commands to the keyboard in the following format:

Øxxx xxxØ – Turn the bell off.
Øxxx xxx1 – Turn the bell on.
1Øxx abcd – Turn on L.E.D. # abcd.
11xx abcd – Turn off L.E.D. # abcd.

The lower-case "x's" are an abbreviation for "don't-care". The corresponding bits may be either Ø or 1.

We will be using the following I/O devices:

BELL Output only. Write a "1" to bit Ø to turn the bell on. Write a "Ø" to bit Ø to turn it off.

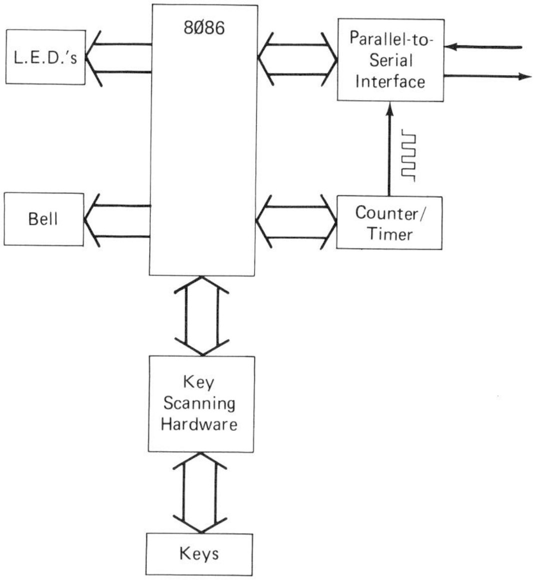

Figure 5.2

LED — Output only. Write "1xxx abcd" to turn on LED # abcd. Write "Øxxx abcd" to turn off LED # abcd.

PSIN — Input only. Read the next byte from the Parallel-to-Serial Interface.

PSSTAT — Input only. Read "ØØØØ ØØØ1" if the transmitter of the Parallel-to-Serial Interface is empty (i.e. ready to transmit). Read "ØØØØ ØØØØ" if the transmitter is not empty.

PSOUT — Output only. Any byte written to this address is transmitted out the Parallel-to-Serial Interface. The transmitter should be empty before writing to PSOUT.

KEYAD — Input only. After an interrupt from the keyboard-scanning hardware, the Key Address of the key which caused the interrupt is read in bits 6:Ø. Bit 7 is unpredictable.

KEYSTATE — After an interrupt from the keyboard-scanning hardware, read "1111 1111" if the key went down or "ØØØØ ØØØØ" if the key went up.

Of course, these are just mnemonic names and not actual device addresses. When we develop an assembly language code a little later on, we will assume that the names listed above have already been defined with the "EQU" pseudo-opcode.

As for the Counter/Timer, we will assume that one channel has already been initialized to provide baud rate timing. Another channel will be used for repeat-key timing. We will assume that there is a subroutine named TIMER_ON which restarts the timer, overriding any current timing operation. TIMER_ON enters with AL equal to the number of milliseconds to be timed; after that many milliseconds, the Counter/Timer will interrupt the 8Ø86. Interrupts will continue to occur at the specified interval until a subroutine named TIMER_OFF is called. TIMER_OFF disables the timing channel of the Counter/Timer.

5.2 Developing a Data Flow Diagram

It happens that the Data Flow Diagram for our keyboard program can be drawn in two unconnected portions. In this section we will develop these portions.

First, let us look at that part of the program which deals with keys. We know that the key-scanning hardware generates interrupts, and that the interrupt routine must read in key addresses and key states. Let's give the name "KBD_INT" to the keyboard interrupt routine. We know that part of the Data Flow Diagram must look like Fig. 5.3.

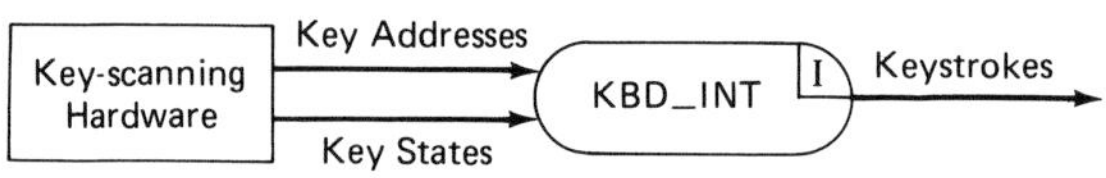

Figure 5.3

The little "I" in the upper right-hand corner of KBD_INT's bubble is a convenient notation to denote that the process being symbolized is an interrupt routine. We see that KBD_INT joins together Key Addresses and Key States to make Keystrokes.

At some point an output driver routine (we'll call it OUT_DRIVER) has to send Keystrokes to the transmitter portion of the Parallel-to-Serial Interface. Keystrokes should only be sent when the transmitter is empty, so OUT_DRIVER has to monitor the status of the transmitter. We have already mentioned that this status is available on input device PSSTAT.

So on the other end of the Data Flow Diagram we need the construction shown in Fig. 5.4.

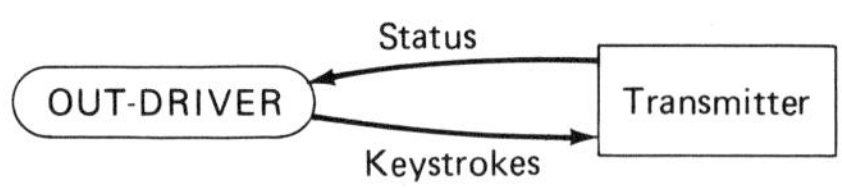

Figure 5.4

Now the question is, how to join up these two parts? We have two alternatives. KBD_INT can either pass Keystrokes directly to OUT_DRIVER, or it can store them in a queue for later processing by OUT_DRIVER. The first solution can be diagrammed like this:

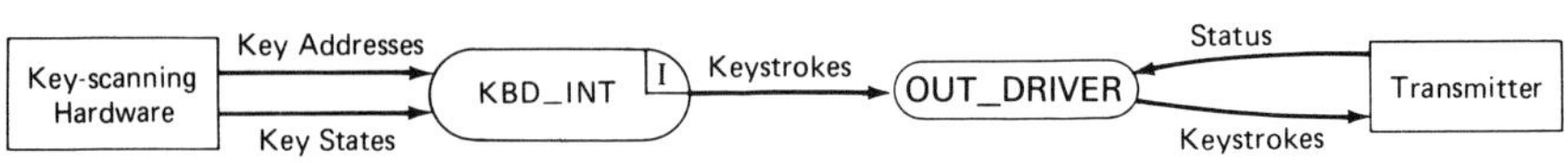

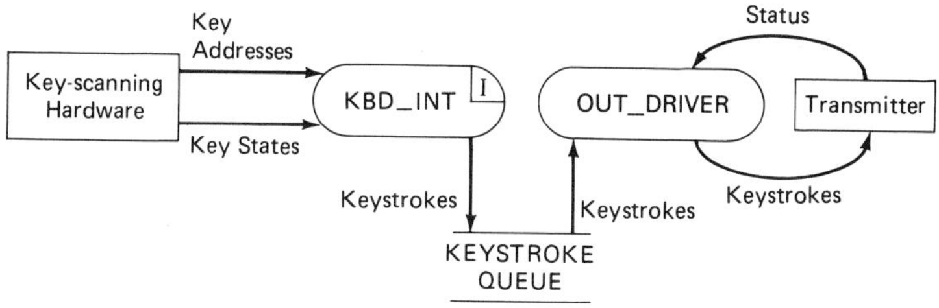

Figure 5.5

Is this a viable solution? Unfortunately it is not. Remember that key strokes can only be transmitted when the transmitter is empty. At ordinary baud rates, it could take a matter of milliseconds to transmit a byte. If the user should type faster than the Parallel-to-Serial Interface can transmit, then the program would fail to notice some key actions. Thus it is necessary for the interrupt routine KBD_INT to store keystrokes in a circular queue, as depicted in the Data Flow Diagram in Fig. 5.5.

On the average, the keyboard will run considerably slower than the transmitter. However, we must be prepared for any eventuality. A good typist can type a short familiar word with little more than a single millisecond between keys.

This is a specific case of a general rule for real-time software design:

- If the input is always slower than the processing or the output, then you don't need a queue.
- If the input is sometimes faster than the processing or then output, then you need a queue.
- If the input is always faster than the processing or the output, the the system can't possibly work. Any queue, no matter how large, will eventually overflow.

Keystrokes may be generated not only by the keys, but also by the repeat-key mechanism. Somehow—and since we are in the design phase we don't really care how just yet—interrupts from the Counter/Timer have to get turned into Keystrokes. The same rule we just quoted motivates us to put Keystrokes generated by Counter/Timer interrupts into the KEYSTROKE QUEUE rather than transmitting them immediately.

The repeat-key mechanism's contribution to the Data Flow Diagram is shown in Fig. 5.6. When we put the two pieces together, we get Fig. 5.7.

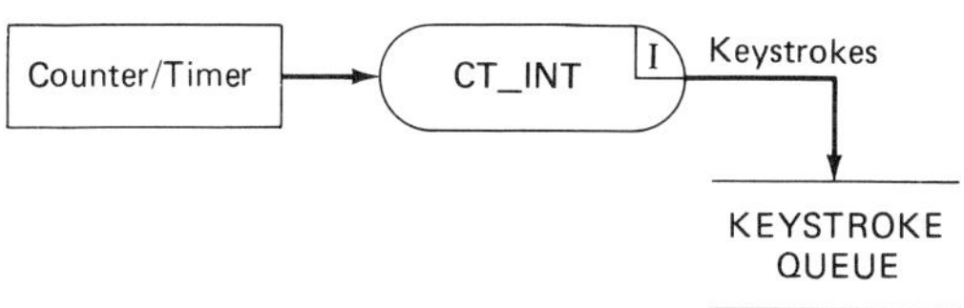

Figure 5.6

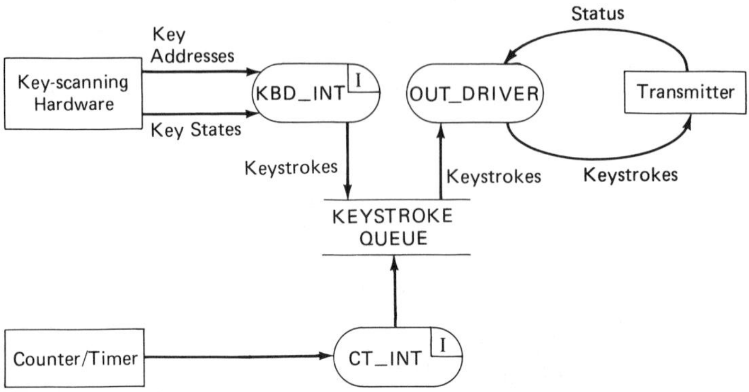

Figure 5.7

We have called the Counter/Timer interrupt routine "CT_INT." Observe that CT_INT receives no data from the Counter/Timer.

The half of the Data Flow Diagram which we have developed so far tells us that we will have to be cautious when we implement the KEYSTROKE QUEUE. It is filled by two interrupt processes and emptied by one background process, and we will have to make sure that these processes do not interfere with one another.

The other half of the Data Flow Diagram involves receiving bell and L.E.D. commands from the host, via the Parallel-to-Serial Interface's receiver. Again, we have the option of immediately processing the input bytes—in this case the commands—or storing them in a queue. The Data Flow Diagrams for the two possibilities are shown in Figs. 5.8 and 5.9.

It will only take a few lines of code to convert raw commands from the system into output bytes for the L.E.D.'s and the bell. In this case the input, which is serial, is always much slower than the output, so there is no need for a queue. The serial receiving interrupt routine, which we have called "RX_INT," will call a command interpreter, which we have called "CMD_INTERP," during the interrupt process. This can be illustrated by the Structure Chart in Fig. 5.1Ø.

Our completed Data Flow Diagram is shown in Fig. 5.11.

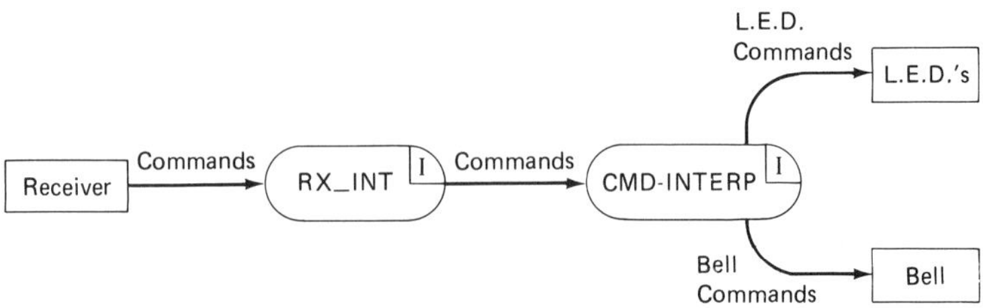

Figure 5.8 No queue.

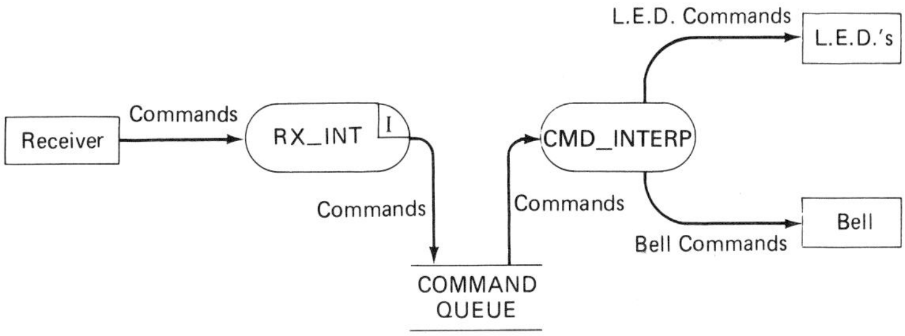

Figure 5.9 With a queue.

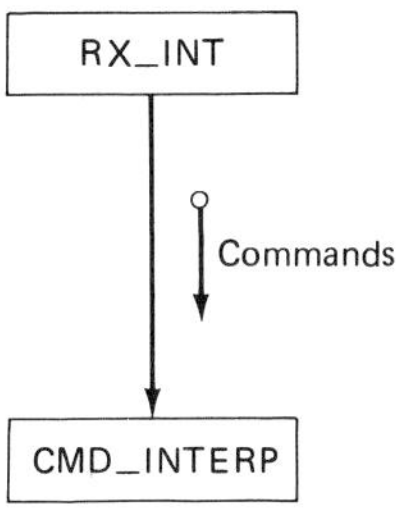

Figure 5.10

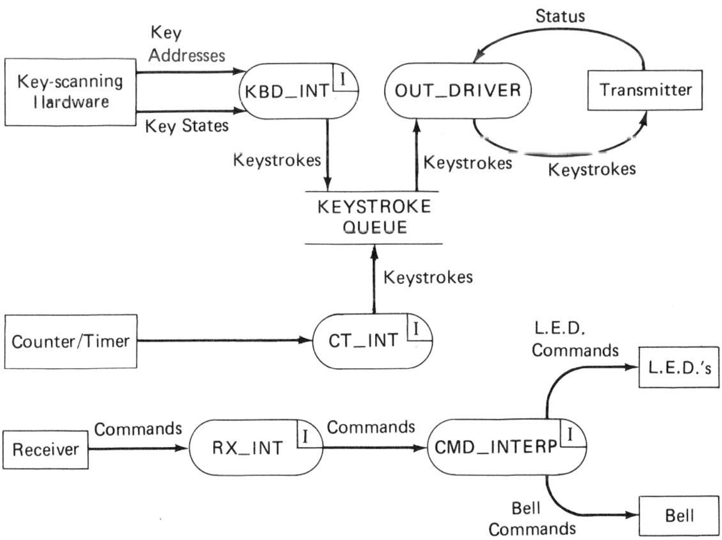

Figure 5.11

5.3 A State Diagram for Repeating Keys

When is it appropriate to stop designing and start implementing? The purpose of design is to bridge the gaps between imagination and code, so it is helpful to defer writing codes until many such bridges are in place.

In our example, although we have finished the Data Flow Diagram, there is one remaining design to do which would greatly simplify the programming process. We have been rather vague about the nature of the repeat-key feature. It is now time to decide how to implement it.

To reiterate, when a key has been held down for 1/4 of a second, reports that it has gone down should be transmitted every 1/Ø of a second. There seem to be three states involved:

STATE #1: When no keys are depressed.
STATE #2: When a key has been down for less than 1/4 of a second.
STATE #3: When a key has been down for more than 1/4 of a second, and has begun repeating.

To ease implementation of this feature, we will develop a State Diagram.

To begin with, what causes a transition out of State #1, and what actions should be taken? The only possible exit out of this state is when a key goes down. At this time the Keystroke should be computed and written into the KEYSTROKE QUEUE. It should also be stored in a variable somewhere, since it is the byte which will have to be retransmitted to the system should the key be held down long enough. We will store the Keystroke in a variable called REP_STROKE. We also need to start a 1/4 second timer, so that we will know when to start repeating. This is easy: we need only call the ready-made subroutine TIMER_ON, with AL = 25Ø.

What we know so far is expressed as a State Diagram in Fig. 5.12.

What about transitions out of State #2? There are three possibilities.

First, if the key goes back up before 1/4 of a second elapses, we should go back to State #1. In this case the timer should be turned off with a call to TIMER_OFF. This can be diagrammed as shown in Fig. 5.13.

The second possiblity is if another key goes down before the 1/4-second time limit. In this case we should update REP_STROKE with the new Keystroke and restart the 1/4-second timer. Note that this is no different from what we do when a key goes down in State #1 (see Fig. 5.14).

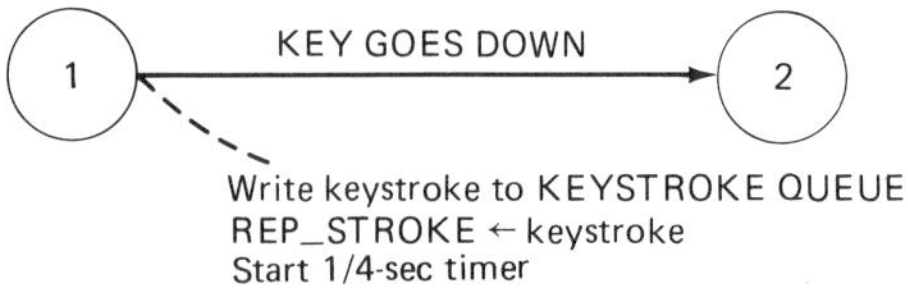

Figure 5.12

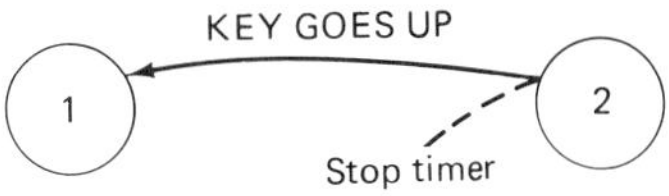

Figure 5.13

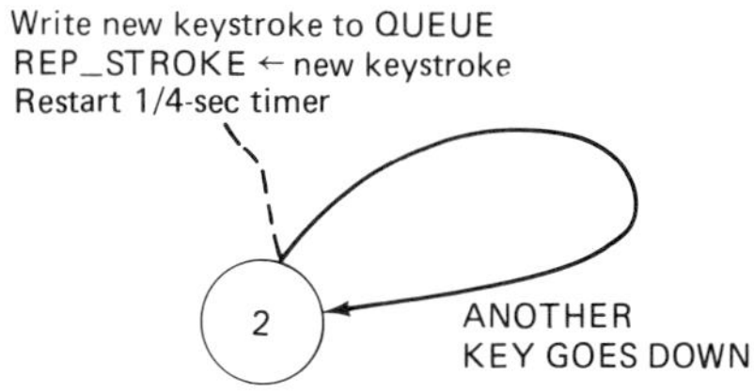

Figure 5.14

The third and most interesting case is when the Counter/Timer interrupts. At this point we should write the keystroke in REP_STROKE to the KEYSTROKE QUEUE, schedule an interrupt in 1/1Ø of a second, and make the transition into State #3 (see Fig. 5.15).

What about transitions out of State #3? This can only occur when the repeating key goes up. As with the corresponding transition out of State #2, the proper action is to turn off the timer (with a call to TIMER_OFF) and switch to State #1 (see Fig.5.16).

If a Counter/Timer interrupt occurs during State #3, we should write REP_STROKE into the queue, restart the 1/10-second timer, and remain in State #3 (see Fig. 5.17).

Putting all the pieces together gives us the composite State Diagram shown in Fig. 5.18.

Based on the State Diagram, we can draw flowcharts for what to do when a key goes down, when a key goes up, and when the Counter/Timer interrupts (see Figs. 5.19–5.21). We will use a variable called STATE to store the current state number.

A glance at these flowcharts reveals that although we faithfully update STATE, we never actually read it or base any decisions on its value. Then do we need it? No! The state of the repeat-key software state machine does not need to be explicitly stored!

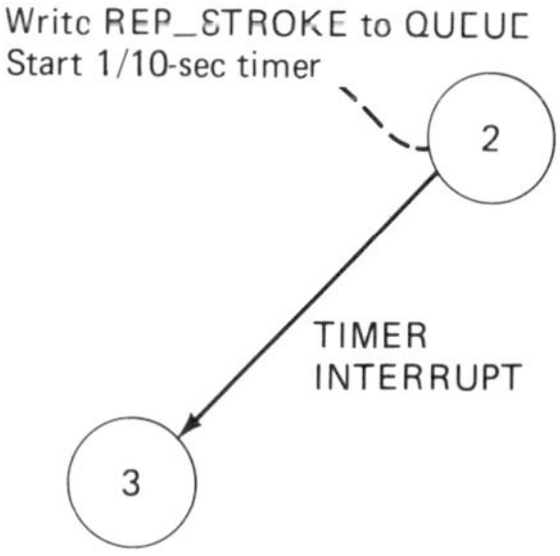

Figure 5.15

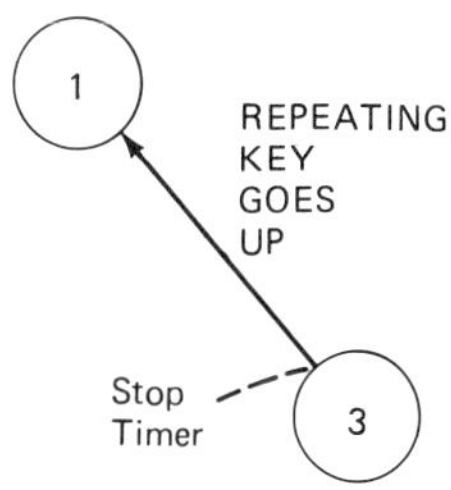

Figure 5.16

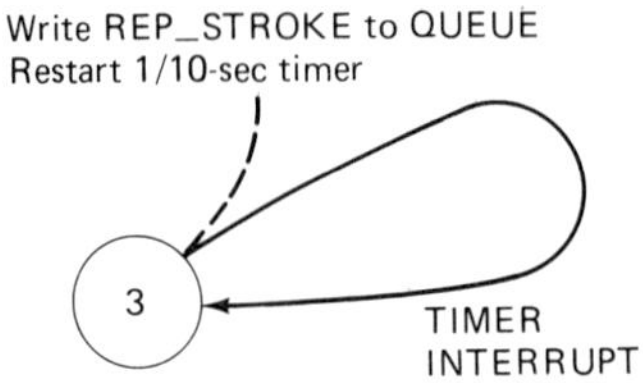

Figure 5.17

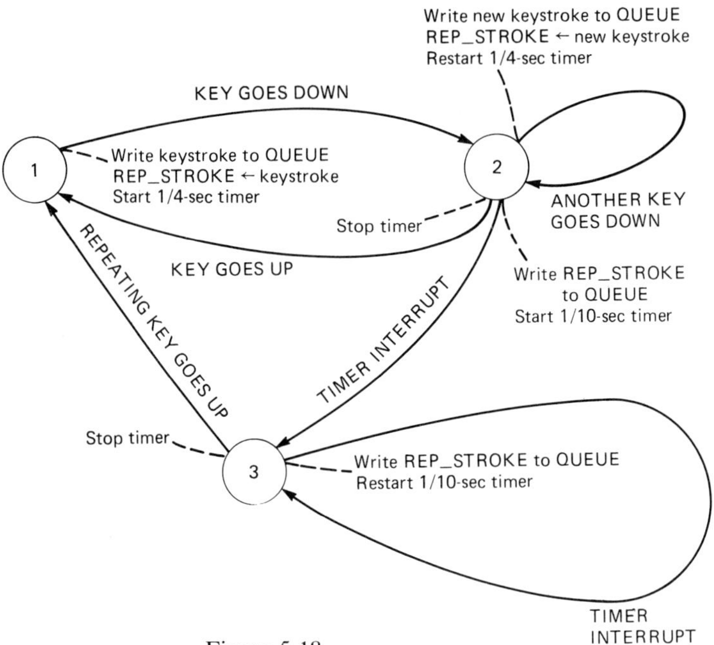

Figure 5.18

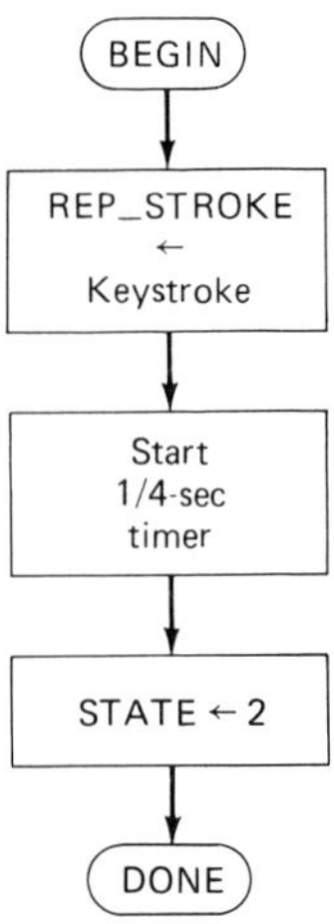

Figure 5.19 When a key goes down.

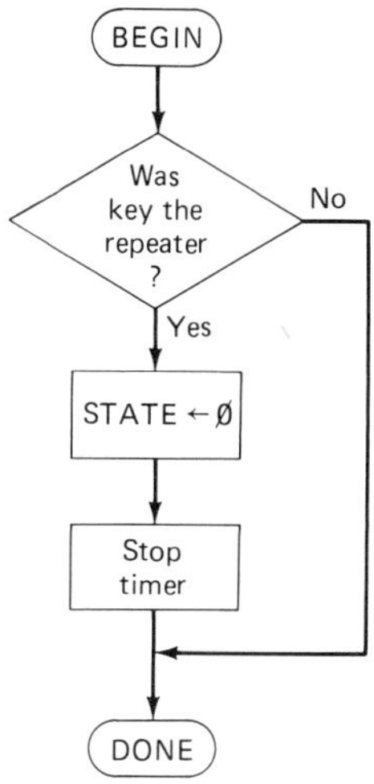

Figure 5.20 When a key goes up.

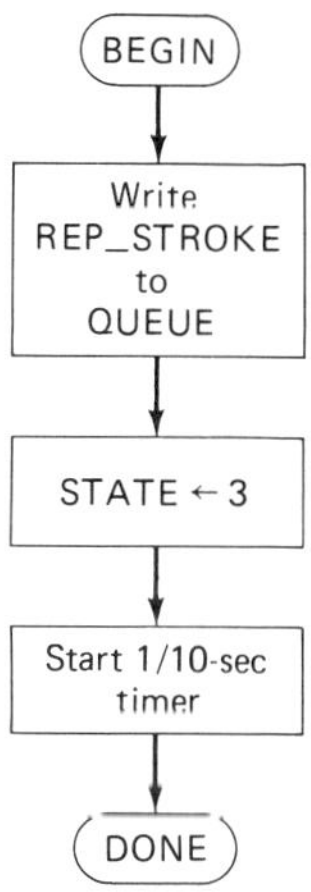

Figure 5.21 When the Counter/Timer interrupts.

5.4 Wherein We Write Some Code

We are ready at long last to write some software. Where should we begin?

This is an important question. A good guideline is to begin with something simple. If you begin with something elaborate and it doesn't work, it will be difficult to know where to start looking for the source of the trouble. The problem might even lie with your prototype hardware and not with your program at all. For every new system you encounter, there is a certain developing-and-debugging expertise which you must acquire, and which can only be acquired by developing and debugging on that particular system. If a sophisticated real-time program is to be written for a certain device, then it is up to you to discover the technology of writing code for that device, and it will be different from the technology of writing code for any other device. If the device has just been designed and nobody has ever written code for it before, then you will be breaking new ground.

When you start writing code on a new system, two things happen at the same time. First, of course, you develop your software. But second, and equally important, you begin learning how to work with the new system. If the system and the program are complicated, it is good to start with the simpler portions of the program, so that you can most efficiently learn about the system.

There is a second consideration for deciding where to start. You will need as much information as possible about what is going on inside the system, so that you will be able to know when it is performing properly and when it has a bug. There are many development tools to help give you this information: oscilloscopes, logic probes, logic analyzers, in-circuit emulators, and resident debuggers, to name a few. If you start with the input end of a path through the Data Flow Diagram and work right through to one of the outputs—especially one of the outputs which you can easily perceive—then you are using the system itself as a source of information about what is going on in the heart of the program.

We assume that you can control the input; when you observe the output, you have created a feedback loop in which you yourself are a participant. Since this is a valuable debugging aid, it is useful to set up some kind of feedback loop as soon as possible. This generally involves implementing the shortest or simplest convenient input-to-output path in the Data Flow Diagram.

Our two guidelines motivate us to begin with that portion of our program which implements bell and L.E.D. commands from the system. The relevant part of the Data Flow Diagram is reproduced in Fig. 5.22. We will write RX_INT and CMD_INTERP. Let's start with the flowcharts in Figs. 5.23 and 5.24.

We ought to write CMD_INTERP first, since in order to write RX_INT

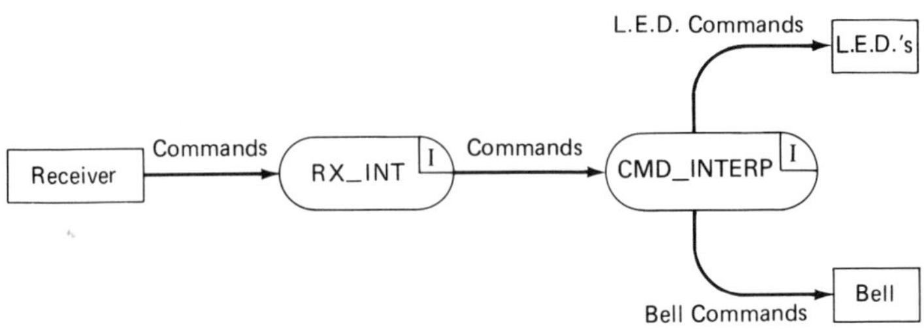

Figure 5.22

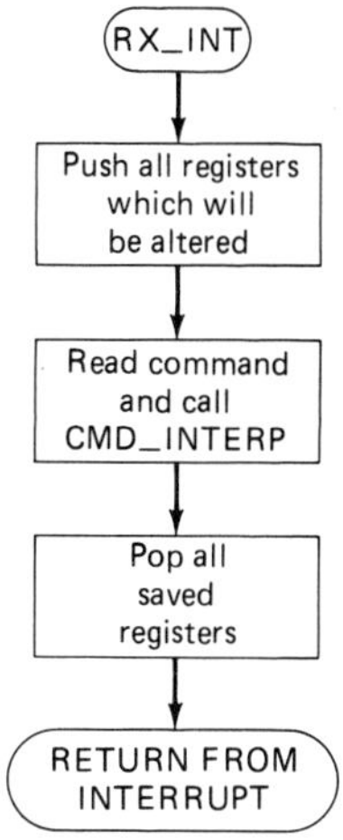

Figure 5.23

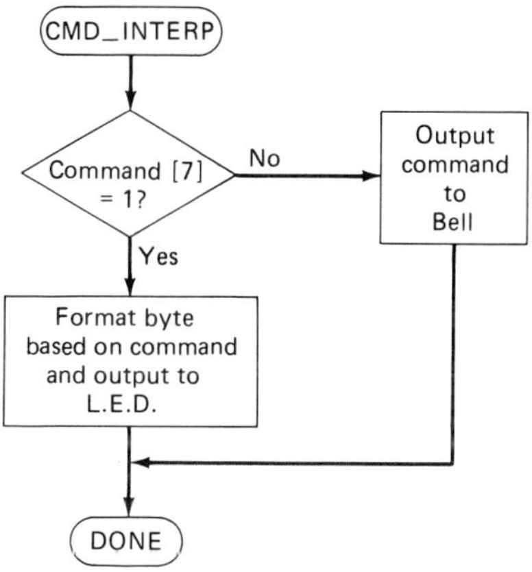

Figure 5.24

we need to know which registers CMD_INTERP will destroy. We will assume that CMD_INTERP enters with the command byte in AL.

```
CMD INTERP:   TEST AL,8ØH       ; BELL COMMAND?
              JNZ  CMD_LED      ; NO, LED COMMAND
              OUT  BELL,AL      ; YES, OUTPUT THE
                                ; COMMAND VERBATIM
              RET
CMD_LED:      MOV  AH,AL
              AND  AH,ØFH       ; AH[3:Ø] SPECIFY
                                ; WHICH LED
```

```
        ADD  AL,AL      ; AL[7] ← 1 FOR ON
                        ; OR Ø FOR OFF
        AND  AL,8ØH
        OR   AL,AH      ; AL[3:Ø] SPECIFY
                        ; WHICH LED
        OUT  LED,AL
```

Now it is easy to write RX_INT:

```
RX_INT: CLI                     ; DISABLE OTHER
                                ; INTERRUPTS
        PUSH AX
        IN   AL,PSIN            ; AL ← COMMAND
        CALL CMD_INTERP         ; EXECUTE THE COMMAND
        POP  AX
        STI                     ; ENABLE INTERRUPTS
        IRET                    ; RETURN FROM
                                ; INTERRUPT PROCESSING
```

Our development philosophy suggests that at this point, were we writing a real program, we should assemble what we have written so far and get it working before we write any more.

Now we need to write the section of the program which deals with the keys themselves. The Data Flow Diagram is reproduced in Fig. 5.25.

Let's start by designing the KEYSTROKE QUEUE and writing its PUT and GET subroutines. Somewhat arbitrarily, we will make the queue 64 bytes long. (Remember that powers of 2 are always the most convenient sizes for circular queues). We need to allocate a block of 64 bytes in data segment RAM; we will call this block "QUEUE." We will also need word variables for the head and tail pointers ("Q_HEAD" and "Q_TAIL") and a byte variable for the number of Keystrokes in the

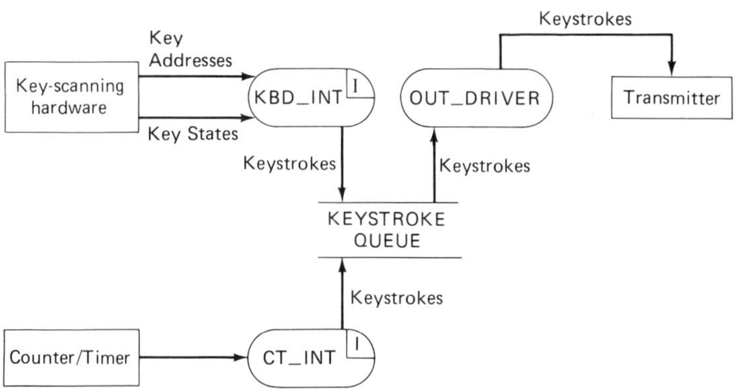

Figure 5.25

queue ("Q_SIZE"). The allocation in the DSEG for all of this storage would look like this:

```
QUEUE     RB    64
Q_HEAD    RW    1
Q_TAIL    RW    1
Q_SIZE    RB    1
```

It is now straightforward to flowchart and write the PUT and GET subroutines. We will use AL to pass keystrokes. If PUT finds the queue full, or if GET finds the queue empty, the Carry Flag will be set (see Figs. 5.26 and 5.27):

```
PUT:          MOV  AH,Q_SIZE
              CMP  AH,64                 ; QUEUE FULL?
              JZ   PUT_FULL              ; YES, ABORT
              INC  AH                    ; NO
              MOV  Q_SIZE,AH             ; INCREMENT SIZE
              MOV  BX,OFFSET QUEUE
              MOV  SI,Q_TAIL
              MOV  [SI+BX],AL            ; STORE
                                         ; KEYSTROKE AT
                                         ; TAIL
              INC  SI
              AND  SI,3FH                ; INCREMENT PTR,
                                         ; MOD 64
              MOV  Q_TAIL,SI
              CLC
              RET                        ; RETURN WITH
                                         ; CARRY CLEAR
PUT_FULL:     STC                        ; QUEUE FULL,
                                         ; RETURN WITH
              RET                        ; CARRY SET

GET:          MOV  AL,Q_SIZE
              CMP  AL,Ø                  ; QUEUE EMPTY?
              JE   GET_EMPTY             ; YES, ABORT
              DEC  BYTE PTR Q_SIZE       ; NO,
                                         ; DECREMENT SIZE
              MOV  BX,OFFSET QUEUE
              MOV  SI,Q_HEAD
              MOV  AL,[SI+BX]            ; AL ← NEXT BYTE
              INC  SI
              AND  SI,3FH                ; UPDATE PTR,
                                         ; MOD 64
              MOV  Q_HEAD,SI
              CLC
```

```
             RET                 ; RETURN WITH
                                 ; CARRY CLEAR
GET_EMPTY:   STC                 ; QUEUE EMPTY
             RET                 ; RETURN WITH
                                 ; CARRY SET
```

Why did we use the instruction DEC BYTE PTR Q_SIZE, rather than the more straightforward DEC AL and MOV Q_SIZE,AL? As it happens, the combination of DEC and MOV, even though it takes two instructions, executes in less time than the single DEC BYTE PTR Q_SIZE instruction. However, we have to take precautions against interrupts. GET runs on the background (i.e., not interrupt) level, and as such is susceptible to being interrupted. If we used the DEC-and-MOV implementation, what would happen if an interrupt came between the DEC and the MOV? Catastrophy!

To see this, suppose that Q_SIZE is 6. After the first line in PUT is executed (MOV AL,Q_SIZE), AL is also 6. The interrupt routine would call PUT, which would increment Q_SIZE from 6 to 7. After the interrupt, GET would proceed to execute MOV Q_SIZE,AL, storing 5 in Q_SIZE. What has happened? Originally there were 6 Keystrokes in the queue. We did one GET and one PUT, so there are still 6 Keystrokes, but

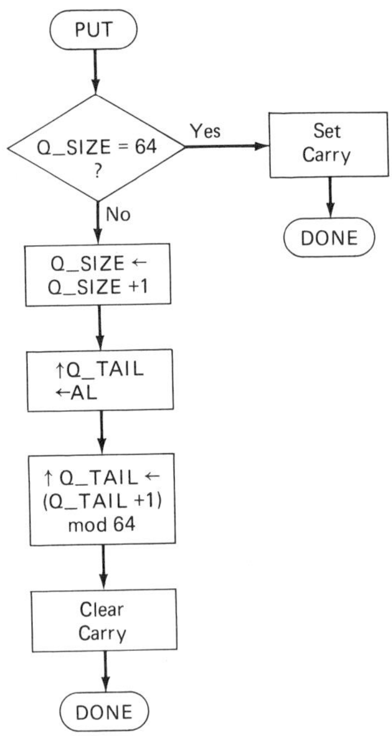

Figure 5.26

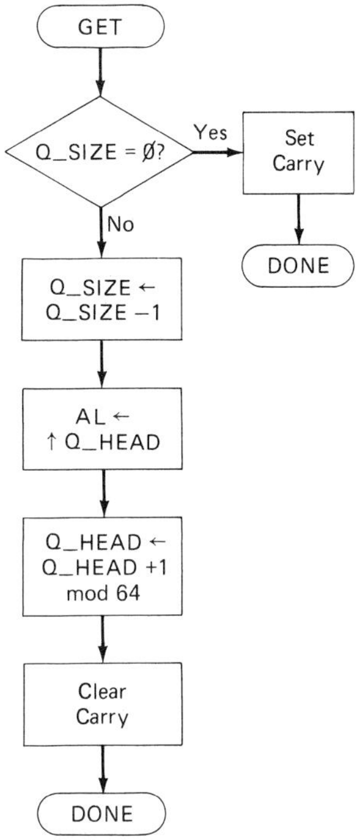

Figure 5.27

Q_SIZE says that there are only 5. Eventually these 5 bytes will be retrieved. Subsequent calls to GET will return with Carry set, indicating that there is nothing left in the queue. The unlucky 6th keystroke will never be processed!

Before we can use PUT and GET, we have to initialize the pointers and Q_SIZE. This must be done during the initialization phase of the program. The code to do it might look like this:

```
MOV AL,Ø
MOV Q_SIZE,AL            ; SIZE ← Ø
MOV AX,OFFSET QUEUE
MOV Q_HEAD,AX            ; HEAD & TAIL PTRS
                         ; POINT TO START
MOV Q_TAIL,AX            ; OF (EMPTY) QUEUE
```

Now that we have the KEYSTROKE QUEUE, let's write OUT_DRIVER, which transmits keystrokes to the system. All that OUT_DRIVER has to do is monitor the transmitter and, when it is empty, check the queue. If there is anything in the queue, the next byte should be retrieved and transmitted (see Fig. 5.28):

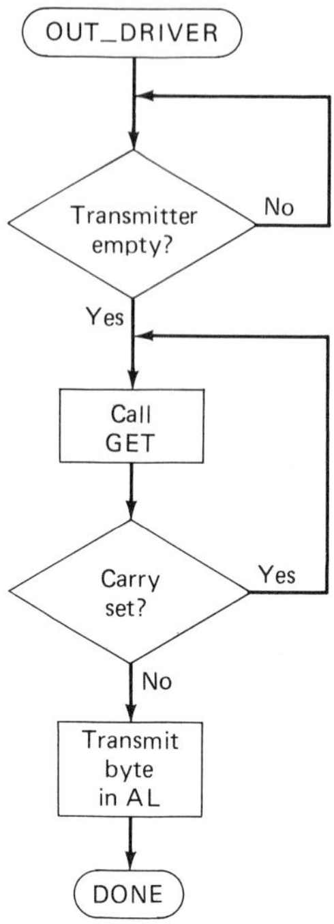

Figure 5.28

```
OUT_DRIVER:     IN    AL,PSSTAT
                AND   AL,1           ; TRANSMITTER
                                     ; EMPTY?
                JZ    OUT_DRIVER     ; NO, WAIT
OUT_DRIVER_1:   CALL  GET            ; YES, ANYTHING
                                     ; IN QUEUE?
                JC    OUT_DRIVER_1   ; NO
                OUT   PSOUT,AL       ; YES, TRANSMIT
                                     ; KEYSTROKE
                RET
```

That was easy! Now what about the key and Counter/Timer interrupt routines? Remember that they are part of the repeat-key software state machine, for which we have already developed a state diagram and flowcharts.

The Counter/Timer interrupt flowchart is shown in Fig. 5.29.

The original flowchart had a "STATE <- 3" box, but you will recall that in the last section we concluded that the variable STATE was unnecessary, so here we have left it out.

The interrupt routine will be called "CT_INT." It will have to save AX and SI, since those registers are destroyed by the call to PUT. We

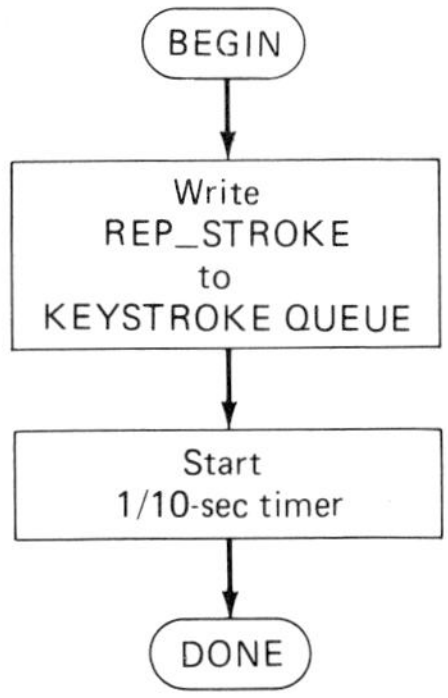

Figure 5.29

assume that TIMER_ON, the ready-made subroutine to start the timer, saves and restores any registers it might use.

```
CT_INT:  CLI
         PUSH AX
         PUSH SI
         MOV  AL,REP_STROKE    ; WRITE REPEATING STROKE
         CALL PUT              ; TO QUEUE
         MOV  AL,1ØØ
         CALL START_TIMER      ; START 1ØØ MSEC TIMER
         POP  SI
         POP  AX
         STI
         IRET
```

We had two flowcharts for interrupts from the keys: one for a key going down, and one for a key going up. We need to combine them. We will call the interrupt routine "KEY INT." The combined flowchart is shown in Fig 5.30.

Once again we have deleted the boxes which deal with the superfluous variable STATE. It is very easy to turn this flowchart into code:

```
KEY_INT:    CLI
            PUSH AX
            PUSH SI
            IN   AL,KEYAD
            AND  AL,7FH
            MOV  AH,AL           ; AH[6:Ø] ←
                                 ; KEY ADDRESS
            IN   AL,KEYSTATE
            AND  AL,8ØH          ; AL[7] ← KEY STATE
            OR   AL,AH           ; AL ← KEYSTROKE
```

```
             CALL PUT             ; PUT KEYSTROKE
                                  ; ON QUEUE
             TEST AL,80H          ; DID KEY GO UP?
             JZ   KEY_INT_UP      ; YES
             MOV  REP_STROKE,AL   ; NO, DOWN
             MOV  AL,250          ; START 250-MSEC
                                  ; TIMER
             CALL TIMER_ON
             POP  SI
             POP  AX
             STI
             IRET                 ; DONE
KEY_INT_UP:  OR   AL,80           ; KEY WENT UP
             CMP  AL,REP_STROKE   ; WAS KEY
                                  ; THE REPEATER?
             JNZ  KEY_INT_Z       ; NO
             CALL TIMER_OFF       ; YES, STOP
                                  ; THE TIMER
KEY_INT_Z:   POP  SI
             POP  AX
             STI
             IRET
```

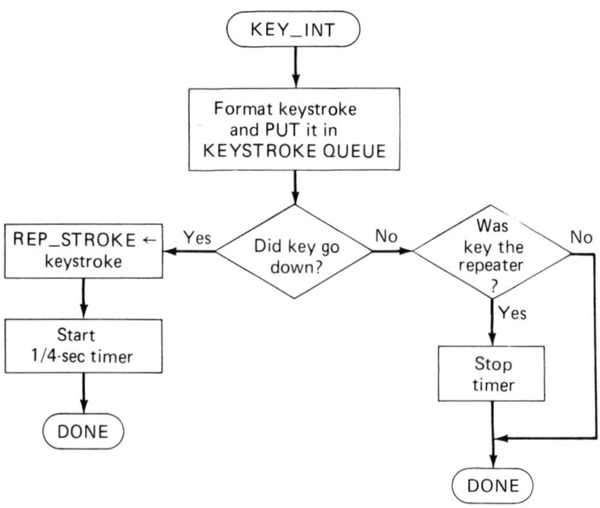

Figure 5.30

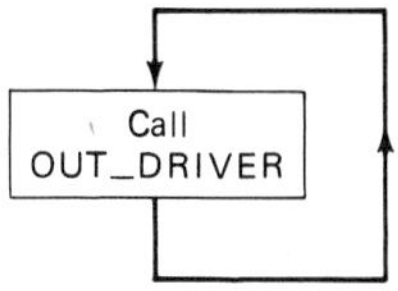

Figure 5.31

All that remains to do is tie the pieces together. We need to arrange for the orderly interaction of our four processes (RX_INT, CT_INT, KEY INT, and OUT_RIVER). This is easy. The first three processes are all interrupt routines. Only OUT_DRIVER is a background routine, so our main background loop is shown in Fig. 5.31.

The program also needs to perform initialization. We have already shown how to initialize the queue. The rest of the initialization involves setting up the hardware, for example the Parallel-to-Serial Interface and the Counter/Timer. We will not write this code here, since it depends on the particular brands of hardware devices chosen by the hardware design engineer.

This concludes our detailed example. We began with a Data Flow Diagram, and also used Flowcharts, State Diagrams, and a little Structure Chart.

While it may have seemed that we devoted an inordinate amount of effort to developing designs, the result was that each portion of code was small, clear, and easy to write. Moreover, the various designs are now available as a road-map through the code for any future engineers who may want to alter the program in any way.

CHAPTER 6
Improving Performance, the Real-Time Railroad, and Goodbye

It is fitting that the last chapter of this book should be about improving performance, because that is typically the last part of developing a program.

One could raise the point that performance improvement, being a development activity, has no place in a book on design. Technically speaking, this is correct. However, if from the outset you design with the knowledge that at the end of it all you will probably have to make improvements somewhere, then those improvements can come naturally and, to a certain extent, painlessly.

Why should you wait until the end of a project to optimize performance? Because if the project is at all complex, it is essentially impossible to know where improvements will be needed. There is a guideline called the "90/10 rule" which states that a typical program spends about 90% of its time executing about 10% of its instructions. The 90/10 rule is only a rule of thumb, and everybody's thumb is different, so it shouldn't be taken too literally. The point of it is that you may be wasting your time if you prematurely streamline the 90% of the program which is rarely executed. Moreover, the more time you spend developing the program and getting familiar with it, the easier it will be to find and correct problems.

6.1 Problems on the Real-Time Railroad

If we think of the Data Flow Diagram as a model for a real-time program, then the problem of improving speed is easy to visualize. The system should be tuned so that data moves as quickly as possible from input boxes and queues, via process bubbles, to queues and output boxes.

To make it even easier to visualize, think of the Data Flow Diagram as a railroad network. Data is represented by the passengers. The stations where they board are the input boxes. The stations where they disembark are the output boxes. Sometimes passengers have to get off a train at a transfer station and wait in line for a second train. These transfer stations correspond to queues. (Nobody enters or leaves the network at the transfer stations.) The process bubbles are of course the trains themselves, picking up passengers at stations and delivering them to other stations.

As with all public transportation systems, everybody complains about how slowly it runs. To make matters worse, only one train is ever running at any one time, because we are modelling a system with only one processor.

The passengers are likely to complain about two different kinds of problems:

1) The train runs too slowly.
2) The train moves quickly enough once it comes, but I have to wait too long at the station.

Of course, the two problems can be combined. We will deal with them one at a time in the next two sections.

6.2 How to Run More Quickly

If a train runs too slowly, it should be speeded up.

Similarly, if a process runs too slowly, it should be streamlined.

Streamlining involves isolating a fragment of code and rewriting it so that it runs faster. The most effective way to do this, and to know whether you have made an improvement, is to calculate the execution times of the old and new versions. This is tedious, as you have to make a calculation for each instruction, but it produces the best results. Every microprocessor's technical manual lists the execution time for each instruction, in units of the clock period.

Proper streamlining strategy varies from processor to processor, and good technique comes with practice. We can, however, offer some general suggestions which apply to all processors.

Store working data in registers, rather than in memory. It *always* takes longer to move data from a register into memory than from a register into another register. Following this guideline provides a double time saving: once when the data is stored, again when it is retrieved.

Store working data in registers, rather than on the stack. This is actually the same principle as the last one, because the stack is part of memory.

Replace subroutine calls with in-line code. In other words, replace statements of the form "CALL ABC" with a copy of subroutine ABC, minus the RETURN instruction. This eliminates a call and a return, both of which are fairly time-consuming.

It is common practice, and generally it is good practice, to "factor" lengthy routines into constituent subroutines, each of which is cohesive and readable. Subroutines are most readable when they fit on a single page. When subroutine calls are replaced by in-line code, this readability is greatly diminished. The situation can be improved somewhat by adding copious artuculate comments.

Speed up queue storage and retrieval. This corresponds to getting passengers on and off the train as quickly as possible,

For a GET routine, there are two cases: when the queue is occupied, and when it is empty. A process which calls a GET routine typically begins as shown in Fig. 6.1.

Speeding up the GET when there is something in the queue (as represented by the "yes" branch in the flowchart) benefits the current process, since it gets its data faster. Speeding up the GET when the queue is empty (the "no" branch) benefits all the other processes in the system, since the current process terminates sooner. This principle also applies to the PUT subroutine.

With a pipelined processor, avoid flushing the pipe. Remember that the 8Ø86 has a pipelined architecture. Part of the machine, the Bus Interface Unit, loads instructions from the Code Segment into a pipeline on board the processor chip. Another part of the machine, the Execution Unit, fetches instructions one by one from the pipeline and executes them.

When program control branches due to a JUMP, CALL, or RETURN instruction, the instructions in the pipeline are no longer valid, since they are not the next few instructions to be executed. In this case the pipeline is flushed by the processor. Since a speed advantage is derived from pipelining, flushing the pipeline is undesirable. The worst case is when branch instructions are regularly distributed, so that the pipeline is always flushed just when it is almost full. (The more nearly full the pipeline is,

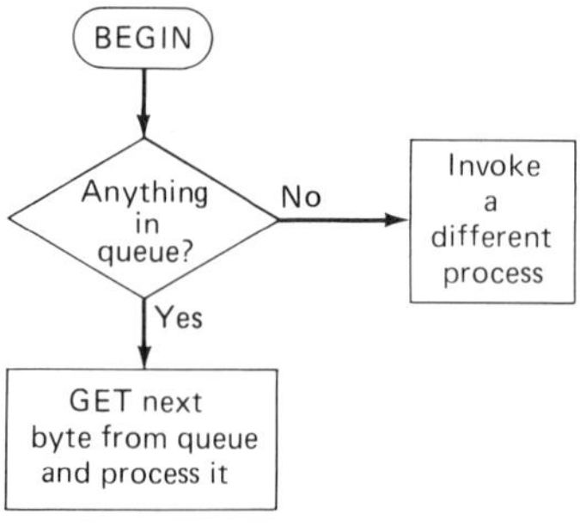

Figure 6.1

the better the processor's performance.) The best case is when nonbranch instructions are grouped together, forming a sort of "straightaway" through which the processor can speed.

When a pipelined processor has to read or write with memory or I/O, the Bus Interface Unit has to stop prefetching instructions and perform the read or the write. This doesn't flush the queue, but is suspends enjoyment of the benefits of prefetched instructions. Again, if the order of instructions can be altered so that there is a "straightaway" of nonbranch, nonread, nonwrite instructions, then optimum use is made of the pipelined architecture. (What's left when you take away branches, reads, and writes? Only interregister instructions, which are always the fastest ones for any processor.)

When doing a conditional jump, jump on the less likely or less critical condition. In most processors, a jump-on-condition instruction takes longer to execute if it has to jump than if it doesn't have to jump.

Suppose we want to test the contents of the CL register and, if it is zero, execute some code named "CLZERO". We could write the test like this:

```
            OR  CL,CL      ; IS CL ZERO?
            JZ  CLZERO     ; YES
CLNOTZERO:  etc, etc
```

If CL is usually zero, then we will usually take the lengthier jump to CLZERO. We can improve the performance of the test by rewriting it like this:

```
        OR  CL,CL           ; IS CL ZERO?
        JNZ CLNOTZERO       ; NO
CLZERO: etc, etc            ; YES
```

The new version will usually fall through to CLZERO, with a significant saving in execution time.

In the language of our railroad analogy, we have listed some guidelines for getting the trains to run faster. Now we will look at getting them to run more frequently.

6.3 How to Run More Frequently

Bear in mind that there is really only one train on our allegorical railroad system. (The train moves quickly enough to create the illusion of multiple virtual trains, like Carmen Miranda at the Copacabana club.) So one way to get the single train to run more frequently on any particular line is to make it run faster on all the other lines, so that it can come back sooner to the line in question. In other words, a process can be made to run more frequently by applying the streamlining techniques discussed in the last section to all of the other processes in the system.

In the rest of this section we will look at a few other methods for getting a train/process to run more frequently.

Run important processes twice in each iteration of the main loop. Suppose we have a system with four processes. The usual implementation would look like Fig. 6.2.

If one process, say Process #3, needs to run twice as frequently, a simple adjustment can be made (see Fig. 6.3).

Run important processes many times in each iteration of the main loop. If keeping the system balanced requires running Process #3, say, twenty times, the adjustment is again simple (see Fig. 6.4).

When a queue is fairly full, run the process which empties it until the queue is fairly empty. In the last example, with Process #3 running twenty times, a lot of time could be wasted if the queue which Process #3 drains is empty. We would find out that it was empty not once, but twenty times, and no other process could run while we were finding out.

It is good to spend time on high-priority processes, but only when there is something to process. The flowchart shown in Fig. 6.5 depicts a scheme for giving a process high priority when its queue is 3/4 full, and maintaining that priority until the queue is 1/4 full.

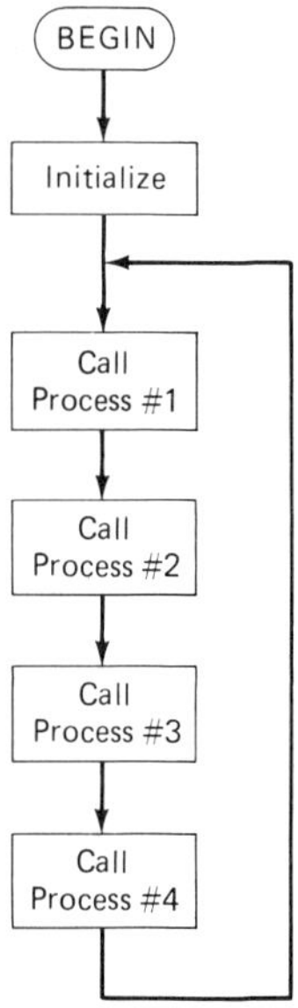

Figure 6.2

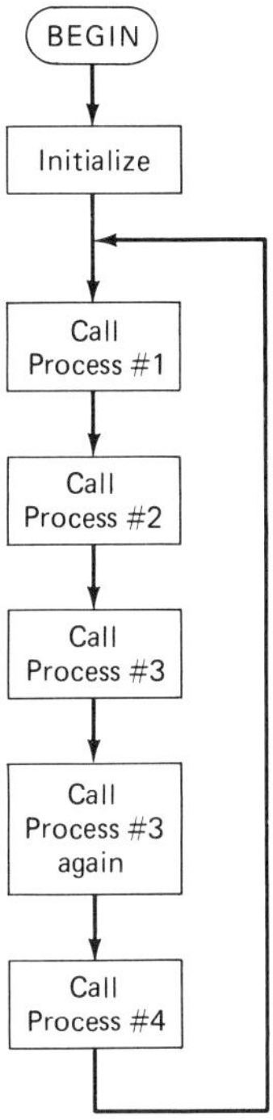

Figure 6.3

This strategy is quite successful in many contexts. The motivation behind it is the idea that when a queue gets mostly full, it is important to drain it until it is mostly empty.

More creative strategies are possible. In general, the priority of a process—that is, the number of times it is called in one iteration of the main loop—can be a function of any or all of the variables which determine the state of the system. The most applicable variables are probably the queue sizes. We would caution the reader against getting too imaginitive, however. The more intricate a priority scheme is, the more processing time is taken up implementing the scheme and calculating priorities, and the less is available for the processes themselves.

6.4 And Goodbye

This brings to a close the last chapter of our investigation on real-time design. We began by looking at the nature of real-time systems and the motivation for taking the time to design the programs which run them. We examined the hardware which composes real-time systems and the design languages which describe them. We worked through a detailed problem, and lastly looked at fine tuning and finishing touches.

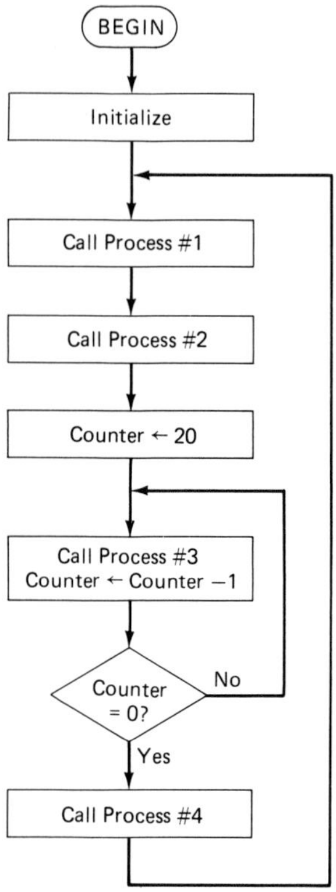

Figure 6.4

As your author, I hope that you have learned something which will make your experience as an engineer more interesting. Let's end with one last word of advice. Today, real-time systems can be found nearly everywhere, from medical equipment to games to weapons. Some products

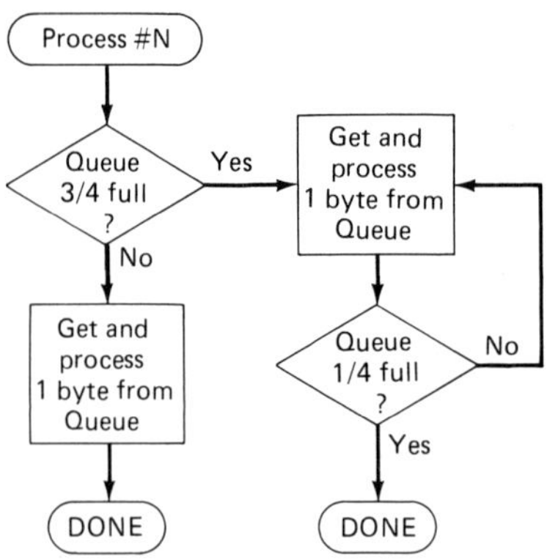

Figure 6.5

create jobs, and some eliminate lives. Some are worthwhile according to one person's standards, and destructive according to another person's. It is important to decide for yourself whether what you create will be positive, negative, or neutral, according to your own personal values. It is my hope that the knowledge you have gained from reading this book will help your professional experience to be as rewarding as possible, according to your own standards. Here ends my short hasty advice. Next textbook is sunk home, and new thoughts intertwine, sharing humanity's advance, nosing toward infinity.

Bibliography

Aho, Alfred v., Hopcroft, John E., and Ullman, Jeffrey D.:*The Design and Analysis of Computer Algorithms*. Reading, Massachusetts: Addison-Wesley, 1974.

Capron, H.L., and Williams, Brian K.: *Computers and Data Processing*. Menlo Park, California: Benjamin/Cummings, 1982.

Clark, K.L., and Cowell, D.F.: *Programs, Machines, and Computation*. New York: McGraw-Hill, 1976.

Hamacher, V. Carl, Vranesic, Zvonko G., and Zaky, Safwat G.: *Computer Organization*. New York: McGraw-Hill, 1978.

Hill, FrederickJ., and Peterson, Gerald R.: *Digital Systems: Hardware Organization and Design. 2nd Edition*. New York: John Wiley and Sons, 1973.

Jantsch, Erich: *The Self-Organizing Universe*. Oxford: Pergamon Press, 1980.

Mano, M. Morris: *Computer System Architecture*. Englewood Cliffs, New Jersey: Prentice-Hall, 1976.

Page-Jones, Meilir: *The Practical Guide to Structured Systems Design*. New York: Yourdon Press, 1980.

Stewart, Mary: *The Crystal Cave*. Greenwich, Connecticut: Fawcett Publications, 1970.

Yourdon, Edward, and Constantine, Larry L.: *Structured Design—Fundamentals of a Discipline of Computer Program and System Design*. Englewood Cliffs, New Jersey: Prentice-Hall, 1979.

Index